NATIONAL ACADEMIES
Sciences
Engineering
Medicine

Strategies to Enable Assured Access to Semiconductors for the Department of Defense

Committee on Global Microelectronics:
Models for the Department of Defense in
Semiconductor Public–Private Partnerships

National Materials and Manufacturing Board

Division on Engineering and Physical Sciences

Board on Science, Technology, and Economic Policy

Policy and Global Affairs

Consensus Study Report

NATIONAL ACADEMIES PRESS 500 Fifth Street, NW Washington, DC 20001

This activity was supported by Contract HQ0034-22-C-0100 with the Department of Defense. Any opinions, findings, conclusions, or recommendations expressed in this publication do not necessarily reflect the views of any organization or agency that provided support for the project.

International Standard Book Number-13: 978-0-309-71702-1
International Standard Book Number-10: 0-309-71702-7
Digital Object Identifier: https://doi.org/10.17226/27624

This publication is available from the National Academies Press, 500 Fifth Street, NW, Keck 360, Washington, DC 20001; (800) 624-6242 or (202) 334-3313; http://www.nap.edu.

Suggested citation: National Academies of Sciences, Engineering, and Medicine. 2024. *Strategies to Enable Assured Access to Semiconductors for the Department of Defense*. Washington, DC: The National Academies Press. https://doi.org/10.17226/27624.

The **National Academy of Sciences** was established in 1863 by an Act of Congress, signed by President Lincoln, as a private, nongovernmental institution to advise the nation on issues related to science and technology. Members are elected by their peers for outstanding contributions to research. Dr. Marcia McNutt is president.

The **National Academy of Engineering** was established in 1964 under the charter of the National Academy of Sciences to bring the practices of engineering to advising the nation. Members are elected by their peers for extraordinary contributions to engineering. Dr. John L. Anderson is president.

The **National Academy of Medicine** (formerly the Institute of Medicine) was established in 1970 under the charter of the National Academy of Sciences to advise the nation on medical and health issues. Members are elected by their peers for distinguished contributions to medicine and health. Dr. Victor J. Dzau is president.

The three Academies work together as the **National Academies of Sciences, Engineering, and Medicine** to provide independent, objective analysis and advice to the nation and conduct other activities to solve complex problems and inform public policy decisions. The National Academies also encourage education and research, recognize outstanding contributions to knowledge, and increase public understanding in matters of science, engineering, and medicine.

Learn more about the National Academies of Sciences, Engineering, and Medicine at **www.nationalacademies.org**.

NOTE: See Appendix F, Disclosure of Unavoidable Conflicts of Interest.

Reviewers

This Consensus Study Report was reviewed in draft form by individuals chosen for their diverse perspectives and technical expertise. The purpose of this independent review is to provide candid and critical comments that will assist the National Academies of Sciences, Engineering, and Medicine in making each published report as sound as possible and to ensure that it meets the institutional standards for quality, objectivity, evidence, and responsiveness to the study charge. The review comments and draft manuscript remain confidential to protect the integrity of the deliberative process.

We thank the following individuals for their review of this report:

LYNDEN ARCHER (NAE), Cornell University
DAVID AWSCHALOM (NAS/NAE), University of Chicago
MEGAN BREWSTER, Impinj
STEPHANIE BUTLER, WattsButler, LLC
AMIT GOYAL (NAE), State University of New York at Buffalo
MELISSA GRUPEN-SHEMANSKY, SEMI
SANDRA HYLAND, Northrop Grumman Corporation
SAMSON JENEKHE (NAE), University of Washington
JUDITH MILLER, Independent Consultant

Although the reviewers listed above provided many constructive comments and suggestions, they were not asked to endorse the conclusions or recommendations of this report nor did they see the final draft before its release. The review of this report was overseen by **JENS-UWE KUHN,** Santa Barbara City College, and **DIANE CHONG (NAE),** Boeing Research and Technology (retired). They were responsible for making certain that an independent examination of this report was carried out in accordance with the standards of the National Academies and that all review comments were carefully considered. Responsibility for the final content rests entirely with the authoring committee and the National Academies.

Acknowledgments

The committee would like to thank the following individuals who added to the members' understanding of the fields of semiconductors and public–private partnerships: Dan Armbrust, Silicon Catalyst; Rob Atkinson, Information Technology and Innovation Foundation; Mike Burkland, Raytheon Missile Systems; Patty Chang-Chien, Boeing Research and Technology; Victoria Coleman, U.S. Air Force; Antonio de la Serna, Siemens Government Technologies; Doug Fuller, Copenhagen Business School; Erik Hadland, Semiconductor Industry Association; Ezra Hall, GlobalFoundries; Dan Hutchison, TechInsights; Mukesh Khare, IBM; Lode Lauwers, IMEC; Mark LaViolette, Deloitte; James Li, BAE Systems; Eric Lin, CHIPS Research and Development Office; Christina Lomasney, Pacific Northwest National Laboratory; Thomas Lopez, Boston Consulting Group; Robert McFarland, Contractor Support for the Department of Defense; Vivek Menon, National Reconnaissance Office; Chris Miller, The Fletcher School at Tufts University; Bora Nikolic, University of California, Berkeley; Ramiro Palma, Boston Consulting Group; Leah Palmer, Mesa Community College, Maricopa Community Colleges; Vanessa Pena, Department of Energy; Jason Rathje, Department of Defense; Melinda Reed, Department of Defense; Jon Rolf, National Security Agency National Information Assurance Partnership; Mark Rosker, Department of Defense; Paul Schaum, National Reconnaissance Office; Tyler Schmidt, Intel Corporation; Brent Segal, Lockheed Martin; Arun Seraphin, National Defense Industry Association's Emerging Technologies Institute; Sujay Shivakumar, Center for Strategic and International Studies; Christie Simons, Deloitte; Alison Smith, Naval Surface Warfare Center-Crane Division; Bryan Smith, Naval Surface Warfare Center-Crane

Division; Duncan Stewart, Deloitte; Neil Thompson, Massachusetts Institute of Technology; Sowmya Venkataramani, Intel Corporation; and Terri Wetteland, Intel Corporation. The committee also reached out to several other semiconductor-related entities who declined the invitation to meet with the committee.

The committee would specifically like to thank those who helped support this study on behalf of the Office of the Under Secretary of Defense for Research and Engineering, including Dev Shenoy, Jayson McDonald, Syd Pope, Daniel Radack, and Estelle McKnight.

Contents

APPENDIXES

Preface

Pursuant a congressionally mandated request, the Department of Defense (DoD) asked the National Academies of Sciences, Engineering, and Medicine to form an ad hoc review committee to assess public–private partnership (PPP) models that have the potential to enable sustainable and resilient production of semiconductor chips in the United States.

DoD requires access to both commercial-off-the-shelf (COTS) semiconductor chips (e.g., those routinely used in computers and electronics) and highly specialized, custom-built chips (e.g., for application in radar systems, high-power density electronics, extreme environments, high-sensitivity sensor systems, and systems requiring higher voltage and frequency ranges than commercial applications). While DoD's chip needs have grown over time as defense technologies have advanced, this growth has been dwarfed by the rapid expansion of commercial semiconductor applications, such that today DoD represents less than 2 percent of the total U.S. market. At the same time, the increased complexity of semiconductor chip architectures has driven up the price for designing and fabricating leading-edge, application-specific integrated circuits, creating price constraints for DoD-specific systems. Finally, the entire semiconductor supply chain has become increasingly globalized over the past three decades, such that the U.S. semiconductor sector today is strongly dependent on other nations, including for defense needs. These intersecting factors have created a challenging environment for DoD, sometimes frustrating its efforts to be nimble and innovative with technologies incorporating semiconductor chips.

This report of the Committee on Global Microelectronics: Models for the Department of Defense in Semiconductor Public–Private Partnerships recommends that the relationship between the commercial and defense semiconductor chip sectors should be strengthened, and that the nation's defense needs can be most efficiently and effectively met by DoD leveraging relationships with commercial manufacturers both for COTS parts and for many, although certainly not all, custom parts. In some sense, the committee advocates for a turn toward an earlier engagement model, operational in the latter part of the 20th century, in which DoD was a valued partner to companies in this sector by supporting new advances, and less an arm's-length chip customer. Overall, DoD needs to become nimbler, adjust its bureaucracy, and move beyond its system of captive fabricators and trusted fabricators toward a more flexible engagement with the sector. Clearly, PPPs can continue to be important vehicles for nurturing these relationships—by providing trusted forums for knowledge sharing and partnership formation. This report cautions, however, that there are already many PPPs operating in the semiconductor space. Rather than DoD launching more PPPs, it would be wise to first make every effort to work within the existing network of PPPs to achieve DoD's strategic objectives. This caution reflects the findings of many prior reports that concluded that successful PPPs take considerable time and careful leadership to establish and to set on the path to productive outcomes, and they also require significant, sustained, multi-decade investments to yield the results the committee seeks in terms of an energized supply chain and a coordinated, coherent technology strategy that yields competitive differentiation.

In executing its charge, the committee met 29 times, with 11 closed sessions and 18 public sessions between May and December 2023, and 8 times, with 5 closed sessions and 3 public sessions in January and February 2024. The committee is particularly grateful to the many people and organizations that have generously provided the information needed to compile this report. A broad array of speakers from government, industry, consultant organizations, nonprofit trade organizations, and academia gave input to the committee. The committee thanks the individuals who made contributions to this study and participated in the committee's meetings.

The committee also thanks the director of the National Materials and Manufacturing Board, Michelle Schwalbe, and senior program officer and study director, Jonlyn (Brystol) B. English, for their help and guidance in performing this study. We also express special appreciation to project staff Erik Svedberg, Amisha Jinandra, Joe Palmer, Sudhir Shenoy, and Mason Klemm for assistance with meeting arrangements.

Liesl Folks, *Chair*
Committee on Global Microelectronics:
Models for the Department of Defense in
Semiconductor Public–Private Partnerships

Summary

OVERVIEW

Microelectronics are a vital part of daily life today, either embedded within the digital devices that people depend on or used to produce, measure, or monitor the non-digital world. Since the first semiconductor integrated circuits were produced nearly simultaneously by Texas Instruments and Fairchild Semiconductor in 1958, firms in the United States have played leadership roles in the global semiconductor industry. Advances in semiconductor technologies have conferred upon the United States incalculable economic and national security benefits. The Department of Defense (DoD) played a critical role in creating the semiconductor industry almost from the very start, funding key innovations and (together with the National Aeronautics and Space Administration [NASA]) acting as a major customer in the early days of the industry. DoD systems have enjoyed technical superiority for decades due in part to the steady cadence of advances in U.S. microelectronics being incorporated into its communications, weapons, and deterrence systems.

Today, however, geopolitical and techno-economic factors have resulted in a hollowing-out of the U.S. semiconductor manufacturing sector, and U.S. semiconductor technology leadership is in question. Fewer and fewer firms can afford the capital expenditures necessary to build and operate the most advanced semiconductor chip fabrication facilities, which now cost tens of billions of dollars, resulting in a concentration of manufacturing capability among a small number of companies. In response to rising costs, many U.S. semiconductor firms have moved their manufacturing and assembly, testing, and packaging facilities overseas, seeking lower costs, reduced operating expenses, robust supply chain ecosystems, and, often, generous support from host nations. Today, the domestic U.S. share of global semiconductor manufacturing value by sales is just 12 percent, down from 37 percent in 1990. World-class contract manufacturers (so-called "foundries") have emerged in Taiwan and South Korea—home today to the leading firms capable of high-volume, leading-edge, logic, and memory chip manufacturing. While U.S. firms remain leaders in important sub-markets like electronic design automation software tools, integrated circuit design, and semiconductor manufacturing equipment, the global semiconductor supply chain has shifted decisively toward countries in the Asia-Pacific region, partly because of supportive government policies and targeted investments in those nations. The bottom line is that U.S. technology leadership in commercial microelectronics is no longer assured, and much of the vast, global supply chain is outside of its span of control.

Against that backdrop, securing access to the very best microelectronics has become a defining challenge for today's DoD. In 2022, DoD released "Microelectronics Vision,"[1] articulating a goal that "DoD will obtain and sustain guaranteed, long-term access to measurably secure microelectronics that enable overmatch, increased operational availability, and support warfighter combat readiness" (p. 1). DoD needs a wide variety of microelectronics to support its missions, ranging from legacy chips required to sustain existing systems, to custom chips with the highest security measures, to cutting-edge chips manufactured with older technology, to state-of-the-art (SOTA) commercial-off-the-shelf (COTS) chips to be embedded in other systems and for the most advanced high-performance computing (HPC) and data centers. Because semiconductor design and SOTA fabrication is extremely expensive today, neither DoD nor its systems suppliers can maintain dedicated SOTA facilities to support DoD's low-volume, high-mix needs; therefore, it is critically dependent on finding ways to access the commercially driven SOTA manufacturing base. DoD's Trusted Foundry (a Program of Record) and Microelectronics Quantifiable Assurance (MQA) programs are two approaches that support a vision for engagement with commercial suppliers, but a comprehensive approach is lacking, and hurdles remain.[2]

In decades past, DoD demand dictated the cadence of innovation in the semiconductor industry, and DoD was seen as an attractive customer. Today, leading semiconductor firms do not seek (and sometimes actively avoid) doing business

---

[1] Department of Defense (DoD), 2022, "Microelectronics Vision," Defense Microelectronics Cross-Functional Team, May, https://media.defense.gov/2022/Jun/15/2003018021/-1/-1/0/department-of-defense-microelectronics-vision.pdf.

[2] MITRE Corporation, 2023, "Microelectronics Quantifiable Assurance (MQA) Independent Assessment," https://www.cto.mil/wp-content/uploads/2023/07/MQA-Assessment-Briefing-for-Release-Distro-A.pdf.

with DoD, focusing instead on lucrative consumer markets. DoD's long timelines for systems development, complex procurement processes, stringent security requirements, export restrictions, silos of expertise, and small volumes make it an unappealing customer for commercial semiconductor firms, which act quickly and decisively to fund and field each generation of technology, engaging in worldwide production for the global consumer market. That is, over a period of decades, there has been a marked shift in the relationship; today, the semiconductor industry no longer seeks support from DoD, but DoD needs the semiconductor industry. DoD will not be first mover at new technology nodes owing to the market complexities and economics of scale. However, it cannot be years (or decades) behind the technology forefront either, which is the direction that DoD concerns over supply chain security have led us toward. DoD must stay close to the early-stage development of new technologies, helping to shape directions and preparing to quickly incorporate new advances into their systems.

The U.S. government, led by the Department of Commerce (DOC), is currently stewarding a once-in-a-generation effort to promote and protect U.S. semiconductor industry leadership, following passage of the CHIPS and Science Act of 2022 (hereafter "the CHIPS Act"). The CHIPS office is in the process of distributing $39 billion of incentives to increase domestic semiconductor manufacturing in the coming months and years. In parallel, in October 2022 and October 2023, DOC imposed broad and novel export controls to restrict China's ability to both purchase and manufacture certain high-end chips critical for military advantage[3]—the latest in ongoing technology protection actions. These efforts to promote and protect the semiconductor industry are reminiscent of U.S. government actions in the 1980s and 1990s, which targeted then-ascendent Japanese semiconductor firms with tariffs and voluntary export restraints and established a variety of public–private partnerships (PPPs), notably SEMATECH and the Semiconductor Research Corporation (SRC), to help sustain U.S. semiconductor industry leadership—efforts that were largely deemed successful. Policymakers today are hoping that history repeats itself. The CHIPS Act calls on all government agencies to partner with industry to protect and promote the U.S. semiconductor industry to ensure economic and national security.

This study was conducted at the request of Congress to address the challenges that DoD is experiencing as it engages with the global microelectronics sector, and to help it better understand how to engage with PPPs to support assured production

---

[3] Bureau of Industry and Security, 2023, "Commerce Strengthens Restrictions on Advanced Computing Semiconductors, Semiconductor Manufacturing Equipment, and Supercomputing Items to Countries of Concern," Press Release, October 17, https://www.bis.doc.gov/index.php/documents/about-bis/newsroom/press-releases/3355-2023-10-17-bis-press-release-acs-and-sme-rules-final-js/file.

and innovation in the semiconductor industry. Specifically, the study addresses four primary questions:

1. What is the competitive position of the United States in the global semiconductor ecosystem? And what are the barriers to sustainable and resilient production of semiconductors in the United States?
2. How can PPPs strengthen semiconductor manufacturing?
3. When partnering with the private sector for semiconductor production, what unique challenges and opportunities exist for DoD to support sustainability and resilience in the U.S. semiconductor ecosystem?
4. What policies for PPPs could be adopted to accelerate the development and adoption of disruptive technologies in the United States that benefit DoD and dual-use needs?

FINDINGS AND RECOMMENDATIONS

The vibrant national discussions regarding CHIPS Act funding allocations and the latter stages of this study occurred largely in parallel. The committee approached the study and the resulting report with a goal of incorporating the concurrent discussions happening in the relevant various communities of practice—among defense policymakers, industry, and analysts, as well as experts in science and technology policy, workforce development, and international policy. The Committee on Global Microelectronics: Models for the Department of Defense in Semiconductor Public–Private Partnerships found that DoD's microelectronics challenges are significant and simply cannot be addressed in isolation. There is a significant risk that the most advanced chips in the world will continue to be produced offshore, as well as by foreign-owned companies, for the foreseeable future—chips that DoD may be unable to do without if it wishes to sustain a competitive stance. As a result, the committee urges DoD to (1) pursue a strategy of sustained and comprehensive engagement with DOC and other agencies to ensure alignment with the CHIPS Act implementation across the government in an effort to reshore and friend-shore these leading-edge capabilities and (2) simultaneously find new ways to engage with commercial semiconductor firms without expecting them to conform with DoD's complex bureaucracy and procurement rules. DoD can benefit greatly from learning to act as a fast follower (Box S-1), nimbly leveraging the commercial success of semiconductor firms in the United States and allied nations, as it seeks to achieve its stated microelectronics vision.

The committee notes that DoD's current efforts to engage with commercial suppliers, which include the Trusted Foundry program and demonstration projects such as Microelectronics Quantifiable Assurance (MQA), Rapid Assured Microelectronics

BOX S-1
Fast Follower

By "fast follower," the committee means that the Department of Defense (DoD) should strive to be a leading-edge customer and the first military user in the world to rapidly adopt, and incorporate into its systems, the newest microelectronic technologies developed by commercial industry. Here the committee is considering mainstream logic, memory, and analog integrated circuit process technologies that are advanced by industry leaders such as Intel, GlobalFoundries, Micron, Texas Instruments, Analog Devices, Nvidia, Qualcomm, Advanced Micro Devices, Inc., and others. DoD cannot hope to match the capabilities of those companies with a captive, defense-unique industrial base, and should instead strive to adopt those technologies for DoD use just as they become available for commercial purposes. DoD may not be the first to ship systems using a new technology node, but it must be the first globally to deploy military systems using that technology (because of militarily advantageous features—lowest energy use, highest performance, smallest size and weight). See also Chapter 6, Principle 5.

Prototypes (RAMP), Rapid Assured Microelectronics Prototypes-Commercial (RAMP-C), State-of-the-Art Heterogeneous Integrated Packaging (SHIP), and SHIP 2.0, are steps in this direction.[4] DoD will need to accept, and mitigate, more risk than it has historically tolerated in these and other new arrangements as it builds the needed substantive partnerships with the commercial sector to secure timely access to the microelectronics technologies it needs to support its missions.

In all its activities, DoD is advised to emphasize long-term (i.e., well beyond 5 years) strategic coordination, investment in emerging technologies, leveraging of commercial advancements, and a modernization strategy that is nimble enough to incorporate emerging technologies and be responsive to global competition. The committee has recommendations related to each of these priorities and areas of emphasis. These recommendations, which emerge throughout this report, are summarized here according to a thematic grouping.

Develop a Comprehensive Department of Defense Microelectronics Strategy

DoD currently lacks an overarching microelectronics strategy, meaning different service branches and program offices are seemingly pursuing uncoordinated efforts related to microelectronics research, development, procurement, sustainment, and modernization. The committee recommends that DoD develop such a strategy with buy-in from relevant stakeholders. The committee notes that a

---

[4] The DoD Trusted Foundry, Microelectronics Quantifiable Assurance (MQA), Rapid Assured Microelectronics Prototypes (RAMP), and State-of-the-Art Heterogeneous Integrated Packaging (SHIP) initiatives are addressed in more detail in the report.

comprehensive strategy does not mean that there is a centralized approach to working with industry but rather that the strategy and its implementation are guided by several distinct pillars, with specific lines of effort under each pillar. (See Recommendation 5.16.)

Develop a 21st-Century Research Agenda for Semiconductors

The committee recommends that DoD prioritize investing in disruptive, sometimes called "leap-ahead," semiconductor technologies[5] as key drivers of future technological advancement. This commitment should encompass both funding of and active involvement in long-term semiconductor research and development, including prototyping and testing, to ensure DoD-specific needs are being addressed. In coordination with the CHIPS interagency agenda, DoD (led by the Office of the Under Secretary of Defense for Acquisition and Sustainment) is encouraged to establish a long-term research agenda focused on post-complementary metal-oxide-semiconductor (post-CMOS) materials and devices. In addition, the committee recommends that DoD accelerate its artificial intelligence and machine learning (AI/ML) implementation strategy as these technologies are driving disruptive applications across many operational fronts and are dependent on access to very large numbers of SOTA commercial chips, which today come from primarily foreign-owned, offshore manufacturing companies.[6] (See Recommendations 5.2, 5.3, and 5.4.)

Develop a Chip Design Initiative

Across the microelectronics sector, the cost to design leading-edge chips has been rising swiftly as the complexity of the designs has increased, challenging the commercial sector and the defense sector alike. The committee recommends that DoD organize and lead a new initiative with industry that would seek to utilize AI/ML tools to substantially accelerate new application-specific integrated circuits design and development, and thus reduce chip design costs.

Furthermore, although DoD does not ordinarily design its own custom chips (relying instead on contractors to do this work), it is important that DoD sustains in-house expertise for modern design to enable it to tap into commercial supply

---

[5] DoD, 2022, "DoD Microelectronics Commons: A National Network for Defense Microelectronics Innovation," Defense Microelectronics Cross-Functional Team, https://www.cto.mil/wp-content/uploads/2022/11/DoD_Microelectronics_Commons.pdf.

[6] DoD, 2023, "Data, Analytics and Artificial Intelligence Adoption Strategy," https://media.defense.gov/2023/Nov/02/2003333300/-1/-1/1/DOD_DATA_ANALYTICS_AI_ADOPTION_STRATEGY.PDF.

chains for manufacturing, and partner effectively with experts in chip design, in order to sustain competitive advantage.

This might be accomplished in several ways. DoD could create a center of excellence in chip design and security that DoD program managers can access as needed. Alternatively, DoD might consider embedding design teams within commercial companies that are producing modern chip designs. By engaging with the skilled design teams in industry, DoD would better understand the design processes that support the most advanced technologies. (See Recommendations 5.5, 5.6, and 5.7.)

The Department of Defense Needs Assured Access to the Most Advanced Chips

Many new and emerging defense technologies, including those which leverage artificial intelligence, machine learning, and data analytics, require very large volumes of the most advanced commercial chips. Most of these chips are currently manufactured offshore and by non-U.S.-owned companies, a fact that prevents DoD use under current policies. To ensure access to these chips, which is vitally important, the committee recommends that DoD develop a long-range strategy, to be supported beyond the timeframe of the CHIPS Act. Within this strategy, the committee recommends that DoD coordinate closely with other government agencies and manufacturing companies and explore ways to link semiconductor capital financing efforts by the U.S. government with agreements and programs to ensure ongoing DoD access to leading-edge chips.

The committee further recommends that DoD collaborate with the National Institute of Standards and Technology (NIST) to ensure that university researchers and start-ups have ready access to advanced facilities for prototyping and scale-up of new chip technologies, which in turn may allow for U.S. chip manufacturers to disrupt the ecosystem and regain clear manufacturing leadership once more. (See Recommendations 5.2 and 5.13.)

Strengthen Private-Sector Engagement

The committee believes that the success of DoD's microelectronics strategy will be largely dictated by DoD's ability to constructively partner with the private sector, especially with leading-edge semiconductor companies, including "fabless" companies (without fabrication plants) that utilize foundries and integrated device manufacturers (IDMs) with their own fabricators. Currently, DoD's microelectronics challenges are exacerbated by a series of self-imposed regulatory and administrative barriers, which make doing business with DoD undesirable in the eyes of many firms, and unnecessarily slow and difficult for those that do. The committee recommends that DoD's microelectronics strategy prioritize private-sector engagement, focusing specifically on the following:

- Reviewing and revising policies that limit DoD from manufacturing custom chips in commercial facilities, and simplifying the procurement processes;
- Developing policies to delegate more decision-making authority to program managers, given inputs from appropriate experts on security risks versus rewards;
- Pursuing export control reforms to reduce the International Traffic in Arms Regulations and Export Administration Regulation barriers to doing business with DoD;
- Continuing development of the Microelectronics Quantifiable Assurance program, which seeks to improve microelectronics sourcing flexibility; and
- Learning how to be a fast follower, to leverage the success and innovation of commercially competitive U.S. semiconductor firms.

(See Recommendations 5.8, 5.9, 5.10, 5.14, and 5.15.)

Engage Robustly with CHIPS Act Interagency Efforts

DoD microelectronics efforts should be aligned with broader interagency CHIPS Act efforts to the extent possible. For example, the committee recommends that DoD's microelectronics workforce initiatives should be coordinated with the CHIPS Act workforce initiatives being stewarded by NIST and the National Science Foundation (NSF). The committee recommends that DoD coordinate with interagency efforts to advance immigration reforms that will increase access to talent, especially those actions recommended by the 2022 President's Council of Advisors on Science and Technology (PCAST) report.[7] DoD's Microelectronics Commons[8] and Next-Generation Microelectronics Manufacturing[9] (NGMM) programs should be closely aligned with the CHIPS Research and Development Office's work, especially that of the National Semiconductor Technology Center

---

[7] President's Council of Advisors on Science and Technology (PCAST), 2022, *Report to the President on Revitalizing the Semiconductor Ecosystem*, White House, September, https://www.whitehouse.gov/wp-content/uploads/2022/09/PCAST_Semiconductors-Report_Sep2022.pdf.

[8] Microelectronic Commons, 2023, "The Microelectronics Commons: A National Network of Prototyping Innovation Hubs," https://microelectronicscommons.org.

[9] DARPA (Defense Advanced Research Projects Agency), 2023, "Next-Generation Microelectronics Manufacturing Aims to Sustain R&D Ecosystem: DARPA Names Eleven Teams to Kick Off Groundbreaking U.S. Chips-of-the-Future Effort," https://www.darpa.mil/news-updates/2023-07-20.

(NSTC).[10] Furthermore, the committee recommends that DoD collaborate closely with CHIPS for America[11] on incentive funding decisions and explore long-term sources of funding that extend beyond the 5-year timeframe stipulated in the CHIPS Act. Finally, DoD should coordinate with the Departments of Commerce, Energy, and State, as appropriate, to explore opportunities for increasing domestic and international semiconductor supply chain resilience. (See Recommendations 4.2, 4.3, 4.8, 5.1, 7.1, and 7.2.)

Address Microelectronics Obsolescence, Modernization, and Supply Chain Resilience

DoD is unique among nearly all other consumers of chips in that it continues to need microelectronics at technology nodes[12] that are decades old to sustain systems that have been in continuous service for that length of time. There is a particular need for approaches that address DoD's unique microelectronics obsolescence challenges and modernization needs. The committee recommends that DoD sustain a concerted focus on supply chain resilience, to include modern design technologies, critical material vulnerabilities, radiation-hardened microelectronics, and advanced packaging. (See Recommendations 4.2, 4.5, 5.15, and 5.16.)

Strengthen Workforce Education Initiatives

The semiconductor sector in the United States will only be as strong as its workforce. DoD has the opportunity to help the nation address the gaps that have emerged in both the science and engineering professional workforce and the technical workforce for semiconductor manufacturing. The committee recommends that DoD work closely with NIST and NSF as they form and implement workforce education programs for the semiconductor sector under the CHIPS Act, as well as form and grow sustainable semiconductor education programs through its own efforts. In addition, DoD should support ongoing efforts to ensure federal government salaries for science, technology, engineering, and mathematics (STEM) employees are sufficient to attract the needed talent to incorporate current and future disruptive technologies, such as those based on AI, ML, quantum information

---

[10] National Institute of Standards and Technology (NIST), "Establishing U.S. Leadership in Future Semiconductor Technologies," National Semiconductor Technology Center, https://www.nist.gov/chips/research-development-programs/national-semiconductor-technology-center.

[11] NIST, "Chips for America," https://www.nist.gov/chips.

[12] In semiconductor manufacturing, a "node" is a measure of the smallest feature size that it is possible to fabricate on the wafer, with the current lithographic tools and processes. Currently, the most advanced logic chip manufacturing is at node dimensions on the order of a few nanometers. DoD today requires some highly advanced chips that are fabricated at 130 nm node or even larger.

systems (QIS), and data science methods, into DoD practices, systems, and architectures.[13] Given the prevalence of non-U.S. nationals in the semiconductor sector, DoD should press for reforms to the immigration system as recommended by PCAST in a 2022 report,[14] including the award of lawful permanent resident status for all individuals with advanced STEM degrees working in semiconductor and other advanced technology areas. (See Recommendations 7.1 and 7.2.)

Adhere to Best Practices for Public–Private Partnerships

The committee reviewed a great many existing and prior PPPs related to semiconductor technologies in which DoD has participated, and generally found that they added value for DoD, especially in the early years of the PPP's operation, presumably when the urgency of the problem was most acute. The committee notes that during the committee's work, DoD has established eight new PPPs related to semiconductor supply, collectively referred to as the Microelectronics Commons, funded for 5 years under the CHIPS Act. Each new PPP requires a robust and sustained investment to attract the needed participants to engage as well as a leadership and administrative structure. *Against that backdrop, the committee did not find any single challenge that could only be solved by creating a specific new PPP, rather than potentially leveraging an existing entity.* If DoD nonetheless determines that one or more new PPPs is advisable to support its semiconductor needs, the committee has several recommendations about its structure and operations.

As for those within the Microelectronics Commons, the PPP should focus on supporting teams to move new technologies from demonstrations of prototypes in research laboratories to validation of components within systems and in relevant environments—the so-called "lab to fab" stages of technology development (i.e., technology readiness levels 4–6[15])—and have substantive industry engagement supported by a mix of public and private funds for an extended period of time. This PPP needs well-articulated goals, metrics for success, and intellectual property (IP) treatment agreed upon by all members at the outset. Regular reviews of the PPP (e.g., quarterly or biannually) should be undertaken.

This PPP should be fully coordinated with the CHIPS for America office to avoid duplication and promote exchange of promising ideas, and this PPP should

---

[13] G. Dille, 2023, "Pentagon Approves Pay Raise for STEM, Cyber Roles at IC Agencies," *MeriTalk*, August 23, https://www.meritalk.com/articles/pentagon-approves-pay-raise-for-stem-cyber-roles-at-ic-agencies.

[14] PCAST, 2022, "Revitalizing the U.S. Semiconductor Ecosystem," https://www.whitehouse.gov/wp-content/uploads/2022/09/PCAST_Semiconductors-Report_Sep2022.pdf.

[15] DoD, 2023, *Technology Readiness Assessment Guidebook*, DoD Office of the Executive Director for Systems Engineering and Architecture, June, https://www.cto.mil/wp-content/uploads/2023/07/TRA-Guide-Jun2023.pdf.

explore partnerships with existing manufacturers to make use of these facilities and to develop technology transition plans at the outset.

Finally, any IP developed by this PPP should be carefully managed based on factors such as technology readiness level and the intended end use of any technologies created within the PPP. For any developments beyond early-stage research (where a consortium model allowing for sharing of information may be appropriate), DoD should recognize private ownership of IP created during the partnership and restrict government ownership narrowly to those extreme situations where chips are being developed solely for use by DoD and where the chips have no commercial application. With this approach, companies will be more willing to fully engage with the PPP. (See Recommendations 4.1, 5.1, and 5.11.)

CONCLUSION

There are a great many challenges facing DoD as it seeks to ensure a supply of microelectronics to both sustain its current capabilities and make certain it has technological advantage over any potential adversaries in the future. The committee is confident that many of these problems are solvable with a mix of targeted investments and changes in policy, process, and practice. However, the committee recognizes that DoD is a large and wide-ranging enterprise, and change is not trivial within such a structure. Among the committee's recommendations, the single most important element is that of building close, trusted, working relationships with the commercial entities that lead innovation in the semiconductor industry to permit DoD to rapidly adopt state-of-the-art technologies, with an urgent focus on artificial intelligence and machine learning technologies which may have transformational impacts in national security spanning many technology domains.

To review the report's structure, Chapters 2 and 3 provide historical background and the foundation for the recommendations that follow in Chapters 4 and 5. Chapter 6 provides a general overview of directions the committee suggests DoD follow. Chapter 7 reviews the workforce education issues in the semiconductor sector, both at professional and technician levels, and policy steps for consideration.

1

Introduction

STUDY BACKGROUND AND STATEMENT OF TASK

The study of the Committee on Global Microelectronics: Models for the Department of Defense in Semiconductor Public–Private Partnerships is a congressionally mandated request authorized in Section 9903(a)(6)(C) of the William M. (Mac) Thornberry National Defense Authorization Act for Fiscal Year 2021 (NDAA FY 2021) for the Department of Defense (DoD) to make recommendations and provide policy options for optimal public–private partnerships (PPPs) and partnership activities. The study was developed jointly between the National Academies of Sciences, Engineering, and Medicine and DoD's Office of the Under Secretary of Defense for Research and Engineering Microelectronics Office in response to the mandate in Section 9903(a)(6)(C) of NDAA FY 2021. The section states that

> In conjunction with the activities carried out under this section, the Secretary of Defense shall enter into an agreement with the National Academies of Sciences, Engineering, and Medicine to undertake a study to make recommendations and provide policy options for optimal public–private partnerships and partnership activities, including an analysis of establishing a semiconductor manufacturing corporation to leverage private sector technical, managerial, and investment expertise, and private capital, as well as an assessment of and response to the industrial policies of other nations to support industries in similar critical technology sectors, and deliver such study to the congressional defense committees not later than October 1, 2022.

Pursuant this, DoD asked the National Academies to convene a consensus study committee to identify, explore, and assess PPP models that have the potential to ensure access for the production of semiconductors in the United States. The committee was assembled with members representing government, academia, and the private sector, which included experts in nanostructured materials and devices; PPPs; science, technology, and innovation public policy and analysis; intellectual property law; economics; international semiconductor policy; national security; and supply chain security, among other areas. The committee reviewed the competitive position of the United States in the global semiconductor ecosystem, how PPPs can strengthen semiconductor manufacturing, unique challenges and opportunities for DoD to support sustainability and resilience of the U.S. semiconductor ecosystem, and policies for PPPs that could be adopted to accelerate the development and adoption of disruptive technologies in the United States that benefit DoD and dual-use needs.

Models for semiconductor PPPs and other strategies are desired to provide understanding of the technical, financial, policy, security, and workforce elements needed to ensure U.S.-based production of semiconductors. Semiconductors are essential components of the devices that make everyday tasks faster and safer, help us attain a healthier and better quality of life, sustain next-generation information and communications systems, secure critical infrastructure, enable national security technologies, and strengthen U.S. competitiveness.

This study is intended to provide guidance for semiconductor PPPs and other strategies for assured supply of semiconductors for the United States, with emphasis on the needs of DoD; the report focuses on the topics outlined in the statement of task. (See Box 1-1.)

BOX 1-1
Statement of Task

A National Academies of Sciences, Engineering, and Medicine–appointed ad hoc committee will identify, explore, and assess public–private partnership models that have the potential to enable assured access for the production of semiconductors in the United States. The committee will produce a consensus report that addresses the following questions.

What is the competitive position of the United States in the global semiconductor ecosystem?

The committee will examine barriers to sustainable and resilient production of semiconductors in the United States and explore what helps drive production and create reliable supply chains of materials, equipment, components, and expertise. This could include an exploration of the industrial policies of other nations in support of industries in similar critical technology sectors.

How can public–private partnerships strengthen semiconductor manufacturing?

Given the inherent strengths and weaknesses within the global microelectronics industry, the committee will explore how to tailor public–private partnership strategies to address different aspects of the supply chain, such as tool manufacturing, fabless design (without fabrication plants), electronic design automation, software development, manufacturing capability and capacity, workforce development, domestic research and engineering capture (e.g., hardware start-ups), and raw materials (e.g., wafers, rare earths). This may include an analysis of establishing a semiconductor manufacturing corporation to leverage private-sector technical, managerial, and investment expertise, as well as private capital.

When partnering with the private sector for semiconductor production, what unique challenges and opportunities exist for the Department of Defense to support sustainability and resilience in the U.S. semiconductor ecosystem?

The committee will examine unique challenges for the Department of Defense in engaging with public–private partnerships on semiconductors. To provide meaningful new insights into the advantages and challenges of public–private partnerships, the committee will consider issues such as the research-design-production feedback loop, intellectual property, workforce development, export controls, global marketplace considerations, and specialized Department of Defense technologies not supported by the commercial industry.

What policies for public–private partnerships could be adopted to accelerate the development and adoption of disruptive technologies in the United States that benefit the Department of Defense and dual-use needs?

Given previously described barriers and challenges, the committee will discuss and recommend approaches for the Department of Defense to drive change. The committee will conduct an assessment of, and response to, the industrial policies of other nations to support industries in similar critical technology sectors, which will include analyses and recommendations for the consideration of U.S.-international partnerships in support of the global marketplace, as well as onshoring efforts. The committee will also examine and describe resources (amounts and types of funding) and actions that could help achieve or maintain a global leadership position for each aspect of the supply chain and effectively leverage private-sector investment.

THE STUDY PROCESS AND DATA GATHERING

The study was conducted over the course of approximately 9 months and consisted of a series of committee meetings and public data-gathering sessions. During this time, the committee held 21 public meetings to gather information from a broad spectrum of stakeholders from across industry, academia, and government to identify, explore, and assess PPP models and other strategies that have the potential to ensure access for the production of semiconductors in the United States. Additional individual research was carried out by members of the committee and staff to inform the report.

THE STRUCTURE OF THE REPORT

The report is divided into eight chapters, including this introductory chapter and a final chapter containing a list of the report's recommendations. Chapter 2 addresses the history of the United States in the global semiconductor ecosystem as well as the current state of the U.S. semiconductor sector compared with other nations. Chapter 3 examines barriers to sustainable and resilient U.S. semiconductor production. These chapters provide the historical background and policy foundations for specific recommendations that follow in the next chapters. The committee's assessment of characteristics needed for effective PPPs can be found in Chapter 4. In Chapter 5, the committee explores the unique challenges and opportunities for DoD in working with the private sector in semiconductor PPPs. Chapter 6 details general principles for DoD to adopt to accelerate the development and adoption of disruptive technologies in the United States that benefit DoD and dual-use needs. Barriers, PPP strategies, and advantages and challenges concerning workforce development for the semiconductor industry are examined in Chapter 7. Chapter 8 provides a list of all the recommendations in the report.

2

The Competitive Position of the United States in the Semiconductor Sector

The committee was asked to review the competitive position of the United States in the global semiconductor ecosystem, including a review of the industrial policies in other nations in semiconductor and related critical technologies. This chapter discusses the history of the United States in the global semiconductor ecosystem as well as the current state of the U.S. semiconductor sector compared with other nations. There is a focus throughout on the Department of Defense's (DoD's) semiconductor needs and its potential role in strengthening the semiconductor sector broadly.

THE POSITION OF THE U.S. SEMICONDUCTOR SECTOR

The current U.S. competitive position in semiconductors has been the recent subject of many thorough studies,[1] so this section will not attempt to duplicate those reports and will instead draw from them. This discussion summarizes the

---

[1] Those reports include the following: W. Hunt, 2022, "Sustaining U.S. Competitiveness in Semiconductor Manufacturing," Center for Security and Emerging Technology (CSET), Georgetown University, January, https://cset.georgetown.edu/publication/sustaining-u-s-competitiveness-in-semiconductor -manufacturing; M. Griffith and S. Goguichvili, 2021, "The U.S. Needs a Sustained, Comprehensive and Cohesive Semiconductor National Security Effort," Woodrow Wilson Center, https://www. wilsoncenter.org/blog-post/us-needs-sustained-comprehensive-and-cohesive-semiconductor- national-security-effort; C. Miller, 2022, *Chip War, The Fight for the World's Most Critical Technology*, New York: Simon and Schuster; S. Kahn, A. Mann, and D. Peterson, 2021, "The Semiconductor Supply Chain: Assessing National Competitiveness, Center for Security and Emerging Technology," January, https://cset.georgetown.edu/publication/the-semiconductor-supply-chain; President's Council of Advisors on Science and Technology (PCAST), 2022, *Report to the President on Revitalizing the Semiconductor Ecosystem*, White House, September, https://www.whitehouse.gov/ wp-content/uploads/2022/09/PCAST_Semiconductors-Report_Sep2022.pdf; Defense Science Board, 2022, *Ensuring Microelectronics Superiority*, https://dsb.cto.mil/wp-content/uploads/dsb/site/ wwwroot/reports/2020s/Microelectronics_ExSumm.pdf; R. Palma, T. Sexton, R. Varadarajan, T. Baker, and A. Bhatia, 2022, "How the US Can Strengthen the Global Semiconductor Ecosystem." Boston Consulting Group," https://www.bcg.com/publications/2022/how-the-us-can-strengthen-the-global- semiconductor-industry; Semiconductor Industry Association (SIA), 2023, *State of the U.S. Semiconductor Industry*, https://www.semiconductors.org/wp-content/uploads/2023/07/SIA_State-of-Industry- Report_2023_Final_072723.pdf; SIA, 2021, *State of the U.S. Semiconductor Industry*, https://www. semiconductors.org/wp-content/uploads/2021/09/2021-SIA-State-of-the-Industry-Report.pdf; A. Varas, R. Varadarajian, J. Goodrich, and F. Yinug, 2020, "Turning the Tide for Semiconductor Manufacturing in the US," SIA and Boston Consulting Group, https://www.semiconductors.org/wp- content/uploads/2020/09/Government-Incentives-and-US-Competitiveness-in-Semiconductor- Manufacturing-Sep-2020.pdf; SIA and Oxford Economics, 2023, *Chipping Away, Assessing and Addressing the Labor Market Gap Facing the U.S. Semiconductor Industry*, July, https://www.semiconductors. org/wp-content/uploads/2023/07/SIA_July2023_ChippingAway_website.pdf; SIA and Boston Consulting Group, 2021, *Strengthening the Global Semiconductor Supply Chain in an Uncertain Era*, https://www.semiconductors.org/wp-content/uploads/2021/05/BCG-x-SIA-Strengthening- the-Global-Semiconductor-Value-Chain-April-2021_1.pdf; Institute for Defense Analysis, 2016, *Semiconductor Industrial Base Focus Study–Final Report*, IDA D-8294, December, https://www. ida.org/-/media/feature/publications/s/se/semiconductor-industrial-base-focus-study—final- report/d-8294.ashx; PCAST, 2017, *Report to the President: Ensuring Long-Term U.S. Leadership in Semiconductors*, January, https://obamawhitehouse.archives.gov/sites/default/files/microsites/ ostp/PCAST/pcast_ensuring_long- term_us_leadership_in_semiconductors.pdf; Office of Science and Technology Policy (OSTP), 2022, "Draft National Strategy for Microelectronics Research," https://www.whitehouse.gov/wp-content/uploads/2022/09/SML-DRAFT-Microlectronics-Strategy- For-Public-Comment.pdf; Department of Commerce (DOC), 2023, *Assessment of the Status of the Microelectronics Industrial Base in the United States*, Bureau of Industry and Security, Office of Technology Evaluation, December, https://www.bis.doc.gov/index.php/documents/ technology-evaluation/3402-section-9904-report-final-20231221/file.

historical role of the federal government in public–private partnerships (PPPs) with the semiconductor sector, the importance of semiconductors as an enabling technology for national security and the overall economy, and the current state of the U.S. semiconductor sector. It discusses the strengths and weaknesses of the overall industrial ecosystem for semiconductors in the United States and the national security implications of semiconductor production in the United States, given current geopolitical considerations. Industrial policies and investments in other nations are noted as well as recent U.S. government actions. Barriers to a resilient domestic semiconductor sector are enumerated with corresponding findings, with a particular focus on DoD's role and needs. Potential elements that could be considered for a DoD semiconductor strategy, including a PPP mechanism, are considered at the end, with recommendations for steps to increase resilience.

This report comes at a similar time as the 2023 release of DoD's first National Defense Industrial Strategy.[2] That document noted that "the events of recent years dramatically exposed serious shortfalls in both domestic manufacturing and international supply chains . . . [demonstrating] America's near wholesale dependency on other nations for many products and materials crucial to modern life." It found that "China became the global industrial powerhouse in many key areas—from shipbuilding to critical minerals to microelectronics. . . . [It] exceeds the capacity of not just the United States, but the combined output of our key European and Asian allies as well."[3] It concluded:

> We need to shift from policies rooted in the 20th century that supported a narrow defense industrial base, capitalized on the DoD as the monopsony power, and promoted either/or tradeoffs between cost, speed, and scale. We need to build a modernized industrial ecosystem that includes the traditional defense contractors—the [Defense Industrial Base] primes and sub-tier defense contractors who provide equipment and services—and also includes innovative new technology developers; academia; research laboratories; technical centers; manufacturing centers of excellence; service providers; government-owned, contractor-operated facilities; and finance streams, especially private equity and venture capital.[4]

This proposed approach highlights broader economic security as an essential foundation for traditional national security. It encompasses a growing concern about microelectronics supply resilience and the growing importance of the semiconductor industry. Accordingly, DoD is currently taking a broader look at the overall capability and capacity of the U.S. semiconductor sector. Of course, this is not the first time DoD has been actively involved in the microelectronics sector.

---

[2] Department of Defense (DoD), 2023, *National Defense Industrial Strategy*, https://www.businessdefense.gov/docs/ndis/2023-NDIS.pdf, p. 8.

[3] Ibid.

[4] DoD. 2023. *National Defense Industrial Strategy*, p. 9.

PRIOR PERIODS OF GOVERNMENT-SUPPORTED SEMICONDUCTOR PUBLIC–PRIVATE PARTNERSHIPS

Semiconductor electronic components[5] have been recognized for at least six decades as a foundational technology for implementation of computer, communications, and information technology systems. Chips are now embedded in virtually every device—they have become as essential to information processing as oil was to energy consumption in the second half of the 20th century, or steel to the first half, only more so. They are essential for defense technologies and systems.[6] By annual global sales, semiconductors are now a U.S. $500 billion industry, and are on a path to become a trillion-dollar industry by 2030.[7] Leading-edge semiconductors are essential for a range of advanced technologies today, as well as for near-future disruptive technologies, particularly artificial intelligence (AI), machine learning (ML), and quantum information systems (QISs),[8] but also for autonomous vehicles and clean energy solutions—all of which have significant national security dimensions. And, of course, they are the core infrastructure hardware component required for a host of societal services, such as the Internet and broadband wireless services, upon which the daily functioning of societies is now based. To sustain the current rate of technology advances, semiconductor sector companies spend more on research and development (R&D) than almost any other sector, reinvesting about one-fifth of revenue annually. U.S.-headquartered companies alone reported spending $60.2 billion on R&D in 2023.[9]

Defining Public–Private Partnerships

The U.S. government is currently launching its third large-scale PPP intervention in this sector. The committee notes that there is no standard, internationally

---

[5] In this context, components include a wide array of devices used in electronic systems, ranging from passive devices, such as transistors and capacitors, to active devices, such as logic chips.

[6] DoD, for example, has contracted with Intel for secure semiconductor foundry services as part of its Rapid Assured Microelectronics Prototypes-Commercial (RAMP-C) program to build up domestic design and production of cutting-edge semiconductor chips. Other firms are also involved in the consortium. D. Sebastian, 2021, "Intel Wins Defense Department Award for Domestic Chip Making," *Wall Street Journal*, August 23, https://www.wsj.com/articles/intel-wins-defense-department-award-for-domestic-chip-making-11629726220.

[7] O. Burkacky, J. Dragon, and N. Lehman, 2022, "The Semiconductor Decade: A Trillion Dollar Industry," https://www.mckinsey.com/industries/semiconductors/our-insights/the-semiconductor-decade-a-trillion-dollar-industry.

[8] See, for example, S.W. Huang, S. Han, W. Lee, et al., 2020, "Symbiosis of Semiconductors, AI and Quantum Computing," IEEE International Electron Devices Meeting (IEDM), 19.1.1–19.1.6, https://ieeexplore.ieee.org/document/9372061.

[9] SIA, 2024, "Industry Factsheet," https://www.semiconductors.org/wp-content/uploads/2024/04/SIA-Industry-Factsheet-2024.pdf.

accepted definition of PPPs. Different nations and agencies have adopted different working definitions. Typically, PPPs involve agreements to provide a public asset or service in which both public and private parties have risks and responsibilities.[10] By sharing risks and capabilities across the sector to achieve a public mission, they can enable outcomes that the public sector alone cannot achieve.[11] Objectives for PPP agreements include such goals as establishing a clear and predictable framework for public and private parties, with specific resource commitments; grounding the agreement in value for money; and using budget transparency to limit risks and ensure the integrity of the agreement.[12] DoD has long utilized PPPs and has even issued guidelines for creating PPPs for product support. This document broadly defines PPPs as contractual collaborations between DoD and nonfederal parties where both parties leverage the expertise, resources, and incentives of the other to achieve mutually agreed-on goals.[13] Mutual benefits are a hallmark, and a broad range of activities beyond product support fit within this framework, from information sharing to R&D to joint development projects. Chapter 4 further elaborates on DoD's use of PPPs.

The First Period: Department of Defense Support for Integrated Circuits

The first DoD semiconductor PPP effort was in the 1950s and 1960s, when military and space users worked with industry to insert advanced, cutting-edge semiconductors into defense and space systems in order to achieve qualitative military superiority for the United States during the Cold War conflict with the former Soviet Union.[14] A rich menu of subsidies for R&D, manufacturing technology

---

[10] World Bank, "About PPPLRC and PPPs," https://ppp.worldbank.org/public-private-partnership/about-us/about-public-private-partnerships, accessed July 30, 2024.

[11] International Monetary Fund, 2001, "Public Private Partnerships," *Finance and Development* 38(3), https://www.imf.org/external/pubs/ft/fandd/2001/09/gerrard.htm.

[12] OECD, 2012, "Principles for Public Governance of Public-Private Partnerships," https://www.oecd.org/governance/budgeting/PPP-Recommendation.pdf.

[13] DoD, 2016, *Public-Private Partnering for Product Support Guidebook*, https://www.acq.osd.mil/log/mr/.policy.html/PPP_for_Product_Support_Guidebook_Oct2016.pdf.

[14] See A. Slomovic, 1988, "Anteing Up: The Government's Role in the Microelectronics Industry," RAND Corporation, https://apps.dtic.mil/sti/pdfs/ADA228267.pdf. Almost 50 percent of semiconductor R&D was paid for by government users over this period (K. Flamm, 1996, "Mismanaged Trade? Strategic Policy in the Semiconductor Industry," Washington: Brookings Institution, p. 36). The federal government's share of the market for discrete semiconductors (individual transistors and diodes) peaked at about 50 percent in 1960 (Flamm, 1996, p. 34; Slomovic, 1988, p. 21; N.J. Asher and L.J. Strom, 1977, "The Role of the Department Defense in the Development of Integrated Circuits," Arlington, VA: Institute for Defense Analysis, https://apps.dtic.mil/sti/tr/pdf/ADA048610.pdf, p. 10). Military and space applications initially accounted for the entirety of the integrated circuit (IC) market when commercial sales of these leading-edge semiconductors commenced in the early 1960s and accounted for the majority of sales through the late 1960s (Flamm, 1996, p. 34; Asher and Strom, 1977, p. 26; Slomovic, 1988, p. 29).

refinement, and new capacity investment was devised, as defense and space systems designers stepped forward and worked with early commercial semiconductor suppliers to provide the initial market for what were initially very costly new electronic components.[15] At this time, DoD played a critical market support role. Integrated circuits were first invented in 1959 and first shipped in 1961. DoD and NASA were almost the only customers for the first 4 years of production.[16] Costs and prices dropped quickly and dramatically, however, stimulated by a rapidly expanding commercial demand for semiconductors in a variety of new applications. As the size of the military market relative to the commercial market declined through the 1970s, military demand started to become less relevant to leading-edge technological innovation in semiconductors, and the close partnerships between DoD and industry faded.

The Second Period: Support for SEMATECH

The second major PPP intervention in the semiconductor sector began in the mid-1980s. Although transistors, integrated circuits, and a host of semiconductor advances were initially developed in the United States, by the mid-1980s, the U.S. semiconductor manufacturing industry was falling behind that of Japan. For one critical process technology, photolithography, the Japanese firms Nikon and Canon had developed technologies that surpassed that of U.S. equipment manufacturers. Japan's semiconductor firms had been working together closely with the relevant materials and equipment manufacturers under the aegis of a collaborative industrial R&D consortia funded by its Ministry of International Trade and Industry (MITI) to achieve this goal. That is, Japan utilized a PPP to gain a leadership position in lithography equipment and processes. Japanese manufacturers also had access to lower-cost financing for investments in leading-edge capacity. By the 1970s and early 1980s, Japan's investments in new manufacturing capacity exceeded that of the United States, and it was clear that the United States was losing its leadership position.[17]

In response, a group of U.S. semiconductor CEOs persuaded the Reagan administration to fund a PPP to restore U.S. semiconductor manufacturing technology

---

[15] Asher and Strom, 1977.

[16] DoD and NASA bought 100 percent of integrated circuit production in 1961 and 1962, 85 percent in 1963 and 1964, and 72 percent in 1965 (see C. Fishman, 2019, "How NASA Gave Birth to Modern Computing," *Fast Company*, June 13, https://www.fastcompany.com/90362753/how-nasa-gave-birth-to-modern-computing-and-gets-no-credit-for-it).

[17] See K. Flamm, 1996, "Mismanaged Trade: Strategic Policy and the Semiconductor Industry," Washington, DC: Brooking Institution Press, c. 10. More generally, on the political economy of this Japanese public–private partnership, see C. Johnson, 1982, *MITI and the Japanese Miracle—The Growth of Industrial Policy, 1925–1975*, Palo Alto, CA: Stanford University Press.

leadership. A public–private consortium, Semiconductor Manufacturing Technology (SEMATECH), was founded in 1987 with 14 leading semiconductor companies.[18] In a reflection of the importance of this new PPP, the first chief executive of SEMATECH to be appointed was Robert Noyce, co-founder of Intel. The Defense Advanced Research Projects Agency (DARPA) administered the government's investment, amounting to $500 million over 5 years, which was matched by the industry participants.[19]

SEMATECH's initial targets were to demonstrate production semiconductor chips with a critical dimension of 0.8 micrometers by 1989, 0.5 micrometers by 1990, and 0.35 micrometers by 1992, all using American-made equipment.[20] By the mid-1990s, these objectives were achieved, and a turnaround in U.S. producers' market share was evident, as shown in Figure 2-1. (Note, however, that this figure displays sales by company headquarters location, not manufacturing location.) At that time, a thoroughly globalized materials and equipment industry was in place, Japan's ascendancy had plateaued, and the U.S. chip industry was seemingly again on sound footing; hence, the SEMATECH restriction to using only American-made equipment for manufacturing seemed increasingly impractical and irrelevant.[21]

In July 1994, the SEMATECH board of directors adopted a resolution thanking the federal government for its partnership, which had successfully spurred industry's turnaround and helped the United States regain world competitiveness. Anticipating the domestic industry's continued good health, the SEMATECH board of directors voted to transfer all of SEMATECH's operations to private funding beginning in 1997. With the end of public sector matching funds, this effectively brought the formal industry–government PPP for SEMATECH to an end. However, SEMATECH sought continuing participation by DARPA and DoD as observers on its board of directors and technical advisory boards, and DoD and

---

[18] L. Berlin, 2005, *The Man Behind the Microchip: Robert Noyce and the Invention of Silicon Valley*, Oxford, UK: Oxford University Press, pp. 257–305; R. Hof, 2011, "Lessons from SEMATECH," MIT Technology Review, July 25, https://www.technologyreview.com/2011/07/25/192832/lessons-from-sematech.

[19] L.D. Browning and J. Shetler, 2000, "SEMATECH: Saving the U.S. Semiconductor Industry," College Station, TX; Texas A&M Press. DARPA received additional funding after SEMATECH's initial 5-year time horizon of 1987–1992, until 1994, for semiconductor R&D at SEMATECH as well as at universities and other entities (General Accounting Office, Federal Research, 1992, "SEMATECH's Technological Progress and Proposed R&D Program," GAO/RCED-92-223BR, July, https://www.gao.gov/assets/rced-92-223br.pdf; General Accounting Office, 1992, "Federal Research, Lessons Learned from SEMATECH," GAO/RCED-92-283, September 11, https://www.gao.gov/assets/rced-92-283.pdf).

[20] J.B. Horrigan, 1999, "Cooperating Competitors: A Comparison of MCC and SEMATECH," Monograph, National Research Council.

[21] Semiconductor Technology Council, 1996, *First Annual Report*, September, https://www.esd.whs.mil/Portals/54/Documents/FOID/Reading%20Room/DARPA/10-F-0709_Semiconductor_Technology_Council_First_Annual_Report_September1996.pdf, pp. 3–13.

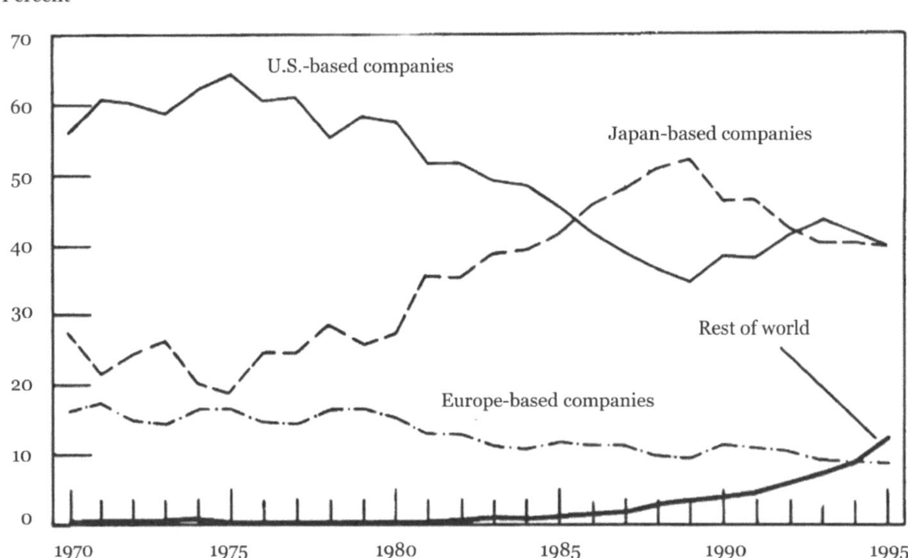

FIGURE 2-1 World market shares of merchant semiconductor companies, by region, 1970–1995. SOURCE: K. Flamm, 1996, *Mismanaged Trade: Strategic Policy and the Semiconductor Industry*, Washington, DC: Brookings Institution Press.

DARPA remained active participants in many ongoing SEMATECH technology-related activities.[22]

---

[22] In its final report to DoD, SEMATECH wrote that "SEMATECH and DoD should continue cooperation beyond 1997. There is no requirement that the existing cooperative framework (e.g., the grant agreement, memorandum of understanding, and enabling legislation) be terminated. Elements of the framework for continued cooperation include but are not limited to DARPA should continue its participation as a non-voting observer on the SEMATECH board of directors and the Executive Technical Advisory Board. DARPA (or a DoD designee of their choice) is encouraged to attend meetings of the focus technical advisory boards and project technical advisory boards. DARPA and SEMATECH should conduct a joint annual review, during which the annual operating and long-range strategic plans of each organization will be shared. (DARPA should share no government classified information with SEMATECH during the course of such reviews.) SEMATECH should continue to be an invited participant in meetings of the Semiconductor Technology Council. Because SEMATECH has ceased to seek federal funds, the possibility of formal SEMATECH representation on the STC should be considered. . . . SEMATECH will continue the use of all assets purchased with regular (commingled) operating funds, unless and until dissolution of the corporation."

Equally notable was SEMATECH's list of elements not to be part of the ongoing cooperation framework: "SEMATECH will no longer be subject to audits by the GAO or other federal audit or oversight agencies. . . . SEMATECH will no longer be subject to the regulations and requirements set forth in OMB circulars A-110, A-122 and A-133." (SEMATECH, 1997, *SEMATECH 1987–1997, A Final Report to the Department of Defense*, February 21, https://www.esd.whs.mil/Portals/54/Documents/FOID/ Reading%20Room/Science_and_Technology/10-F-0709_A_Final_Report_to_the_Department_ of_Defense_February_21_1987.pdf, pp. 12–13).

This decision reflected three realities of that moment.[23] First, renewing federal funding for SEMATECH would have required a new legislative initiative, and the U.S. semiconductor industry was not ready to invest the necessary political capital, given the apparent success already achieved in reestablishing U.S. industry's competitive position. Second, leading-edge semiconductor manufacturing technology was being transformed into a globalized infrastructure, with best-in-class equipment and materials available from friendly foreign sources. The U.S. semiconductor industry leaders supported the globalization tide,[24] and the prevalent view at DoD was that by incorporating global technology leaders into SEMATECH, including suppliers based in allied nations, the U.S. military would be able to ensure U.S. industrial and defense access to the very best available semiconductor technology. The globalization of SEMATECH might also have been seen as problematic from the standpoint of seeking new legislation to provide federal support to SEMATECH. Third, at that time, "about 60 percent of SEMATECH's annual budget was being expended on external R&D efforts,"[25] and both SEMATECH and DoD leadership viewed the National Science Foundation (NSF), DARPA, and the nonprofit Semiconductor Research Corporation as having better capabilities to identify and fund long-range R&D on next-generation semiconductor manufacturing innovation.

In response to these developments, SEMATECH underwent two structural changes in the mid-1990s. First, it became a truly international consortium, reflecting the continuing globalization of the semiconductor manufacturing, equipment, and materials ecosystem. In 1996, "SEMATECH established the International 300 mm Initiative (I300I), a subsidiary organization with international membership . . . a major technology and infrastructure development initiative to lead its membership and their equipment suppliers in the conversion to 300 mm diameter wafers."[26] A few years later, SEMATECH would change its name to International SEMATECH and invite foreign members to join—and join they did.[27]

---

[23] This description is based on the recollections of Kenneth Flamm, who was serving as DoD's top dual-use technology policy official at the time and was charged with interfacing with SEMATECH's CEO and top leadership, and of William B. Bonvillian, who was serving as a senior advisor in the U.S. Senate working on these issues at that time.

[24] The U.S. semiconductor industry had been one of the very first U.S. industries to move the bulk of its most labor-intensive operations (chip assembly and packaging) offshore in the mid-1960s. See K. Flamm, 1985, "Internationalization in the Semiconductor Industry," in J. Grunwald and K. Flamm, *The Global Factory*, Washington, DC: Brookings Institution.

[25] Semiconductor Technology Council, 1996, *First Annual Report*, pp. 3–13.

[26] Semiconductor Technology Council, 1996, *First Annual Report*, pp. 4–5.

[27] By 2006, full international members of SEMATECH included Philips, Infineon, and Samsung. In addition to these firms, TSMC, Panasonic/Matsushita Electric, Spansion, NEC, and Renesas were members of the I300I subsidiary.

Second, SEMATECH shifted its focus to developing semiconductor manufacturing standards and coordinating the technological improvements needed to ensure successful and timely introduction of new semiconductor manufacturing technology "nodes" using products from the many suppliers making up the global semiconductor industrial base. SEMATECH had been "organized into thrust areas that are aligned with the Semiconductor Industry Association's (SIA's) National Technology Roadmap for Semiconductors."[27] The development and commercialization of the next generation of lithography tools and processes (at a wavelength of 193 nm), and the coordinated development of new tools able to use larger-diameter wafers (at 300 mm) were the first R&D thrust areas chosen by SEMATECH following these changes.[28]

SEMATECH's (and the U.S. semiconductor industry's) embrace of industry globalization also included pivoting its stewardship of the National Technology Roadmap for Semiconductors effort to instead developing an International Technology Roadmap for Semiconductors (ITRS), and using that as the vehicle to coordinate and accelerate semiconductor manufacturing technology development on a global scale.[28] SEMATECH became the administrative hub for the ITRS and provided a unique and useful service to a global semiconductor industry in coordinating the development and deployment of new technology nodes through the first half decade of the 21st century.

These new institutional arrangements seemed to work well for SEMATECH through the beginning of the new millennium, but then seemed to flounder. Arguably, the fundamental problem was economics. With rapidly increasing economies of scale at each new generation of manufacturing technology, fewer and fewer firms were willing or able to invest in building expensive semiconductor wafer fabrication plants, known as fabs, that used cutting-edge semiconductor technology. Indeed, leading-edge semiconductor manufacturing was becoming an increasingly concentrated industry. SEMATECH evolved into a technology coordination entity in which only a few of its members were likely to invest in the next generation of manufacturing technology. Having a large group of bystander firms making decisions about technology priorities that would affect a much smaller group of firms that were likely to make the needed investments became an untenable organizational structure. With the industry's recovery, DoD and the federal government in following years had only a limited engagement with the industry and were not prepared to consider future subsidies to offset those deployed by other nations to capture market share. In 2003, SEMATECH entered into a partnership

---

[28] For a more detailed discussion of the ITRS, see K. Flamm, 2009, "Economic Impacts of International R&D Coordination: SEMATECH and the International Technology Roadmap," pp. 108–125 in National Research Council, 2009, *21st Century Innovation Systems for Japan and the United States: Lessons from a Decade of Change: Report of a Symposium*. Washington, DC: The National Academies Press. https://doi.org/10.17226/12194.

with the State of New York's SUNY Polytechnic in Albany, subsequently relocating to Albany, and in 2010 folded its operations into SUNY Poly with some leading members withdrawing, effectively ending the program.[29]

Defense Advanced Research Projects Agency
Support for Semiconductor Advances

Keys to the success of SEMATECH as a PPP during its early years included the following: bringing in top technical talent from industry partners for 1- or 2-year assignments; the personal commitment of industry CEOs; leadership that encouraged innovative thinking; a no-nonsense attitude toward dropping noncompetitive approaches; collaborative development of roadmaps for future industry-wide technology goals; and DARPA's close cooperation, which encouraged close coordination among the private partners.[30]

DoD also made a series of smaller-scale semiconductor research investments with industry and academic researchers in the years before and after SEMATECH. Perhaps most notable was work in the late 1970s by researchers at the California Institute of Technology (Caltech) and Xerox Palo Alto Research Center (PARC) to develop a greatly improved system for Very Large Scale Integration (VLSI) chip design using advanced software tools.[31] Development of these tools was a key enabler for the separation of chip design from chip manufacturing, and the development

---

[29] R. Van Steenburg, 2022, "With the CHIPS Down, Sematech Gets Second Look," *National Defense*, June 2, https://www.nationaldefensemagazine.org/articles/2022/6/2/with-chips-down-sematech-gets-second-look; C. Wessner and T. Howell, 2023, "Implementing the CHIPS Act: SEMATECH's Lessons for the National Semiconductor Technology Center," CSIS, May 19, https://www.csis.org/analysis/implementing-chips-act-sematechs-lessons-national-semiconductor-technology-center; M. Lapels, 2015, "Semiconductor R&D Crisis?" *Semiconductor Engineering*, June 11, https://semiengineering.com/semiconductor-rd-crisis.

[30] See, for example, P. Grindley, D. Mowery, and B. Silverman, 1994, "SEMATECH and Collaborative Research: Lessons in the Design of High-Technology Consortia," *Journal of Policy Analysis and Management*, https://www.academia.edu/15457838; G. Moore, 2003, "The SEMATECH Contribution," pp. 96–104 in National Research Council, 2003, *Securing the Future: Regional and National Programs to Support the Semiconductor Industry*. Washington, DC: The National Academies Press, https://doi.org/10.17226/10677; General Accounting Office, 1992, *Lessons Learned from SEMATECH*, GAO/RECD 92-283, September, https://www.gao.gov/assets/rced-92-283.pdf; A.N. Link, D.J. Teece, and W. Finan, 1996, "Estimating the Benefits from Collaboration: The Case of SEMATECH," *Review of Industrial Organization* 11(5):737–751, https://www.jstor.org/stable/41798663; K. Flamm, 2009, "Economic Impacts of International R&D Coordination: SEMATECH and the International Technology Roadmap," in National Research Council, 2009, *21st Century Innovation Systems for Japan and the United States: Lessons from a Decade of Change: Report of a Symposium*, Washington, DC: The National Academies Press, https://doi.org/10.17226/12194.

[31] See T.S. Perry and P. Wallich, 1985, "Xerox Parc's Engineers on How They Invented the Future," *IEEE Spectrum*, October 1, https://spectrum.ieee.org/xerox-parc.

of fabless (design-only, without fabrication plants) and foundry (manufacturing-only) chip firms followed. This work had been supported by around $100 million in government funding by 1982, via direct investments rather than through a PPP structure.[32]

Xerox PARC's efforts in the late 1970s to offer a fast-turnaround VLSI prototyping service allowed researchers to submit chip designs as digital files uploaded remotely to cooperating chip manufacturers for the first time. This system became the basis for a DARPA-supported chip prototyping service known as Metal Oxide Semiconductor Implementation Service (MOSIS), which continues to serve university and industry researchers, hosted by the University of Southern California.[33]

> Requests from different researchers were pooled into common lots and run through the fabrication process, after which completed chips were returned to the researchers. This system obviated the need for direct access to a fabrication line or for dealing with the complexity of arranging fabrication time at an industrial facility, by providing access to a qualified group of fabrication facilities through a single interface. Prominent VLSI researcher Charles Seitz commented that MOSIS represented the first period since the pioneering work of Eckert and Mauchley on the ENIAC in the late 1940s that universities and small companies had access to state-of-the-art digital technology.[34]

This service effort (which grew from 258 projects in 1981 to 1,880 in 1989)[35] may have provided some of the inspiration, as well as some of the early software infrastructure,[36] for the pure-play foundry business model developed by Taiwan Semiconductor Manufacturing Company (TSMC) when it was founded by Morris Chang in Taiwan in 1987.[37]

---

[32] See Chapter 4, "The Organization of Federal Support: A Historical Review," in National Research Council, 1999, *Funding a Revolution: Government Support for Computing Research*, Washington, DC: National Academy Press. https://doi.org/10.17226/6323.

[33] The MOSIS Service. See https://www.themosisservice.com.

[34] National Research Council, 1999, *Funding a Revolution*, p. 121.

[35] Ibid.

[36] "At the time, the interface between the foundry and the design group was fairly simple. The foundry would produce design rules and SPICE parameters for the designers; the design would be given back to the foundry as a GDSII file and a test program. Basic standard cells were required, and these were available on the open market from companies like Artisan, or some groups would design their own." See D. Nenni, 2023, "The TSMC OIP Backstory," SemiWiki [blog], September 18, https://semiwiki.com/semiconductor-manufacturers/tsmc/335155-the-tsmc-oip-backstory; on DARPA funding of SPICE and other chip design tools, see K. Flamm, 1987, *Targeting the Computer*, Washington, DC: Brookings Institution, p. 69.

[37] For a more complete description of these efforts, see Chapter 4, "The Organization of Federal Support: A Historical Review," in National Research Council, 1999, *Funding a Revolution*.

DARPA also funded advances in semiconductor lithography and in computing, from personal computing to computer graphics.[38] The Very High-Speed Integrated Circuit (VHSIC) program was an R&D effort from 1979–1988 through a DoD–industry partnership featuring joint government–industry planning, large-scale industry participation, and multi-firm collaboration to speed improvements and adoption of submicron integrated circuits. In contrast to the later SEMATECH effort, VHSIC's objective was to incorporate cutting-edge semiconductor technology in defense-unique chips, and then into new systems built by DoD's largest contractors, to thus widen the technological advantage relative to the former Soviet Union. Because the scale of production of chips going into weapons systems was so much smaller than the rapidly expanding commercial marketplace, and because security restrictions hindered incorporation of the technology developed into commercial products, the initiative was terminated in 1990, with the commercial sector having moved ahead of DoD's technologies. Nonetheless, the VHSIC program made some important technological contributions to the semiconductor industry. In particular, in parallel with DARPA's VLSI program, VHSIC funded the development of electronic design automation (EDA) tools and a hardware description language, VHSIC Hardware Description Language (VHDL), which ultimately evolved into the chip design software tools in use today.[39] While falling short of SEMATECH in both scale and outcome, VHSIC was a notable response to a well-publicized effort in Japan by MITI (the VLSI Project), with five major Japanese electronics firms targeting commercial IC markets that proved to be very successful.[40]

DARPA's investments in advanced lithography were similarly closely coordinated with industry. Over a 14-year period from fiscal year (FY) 1991 to FY 2005, DARPA's advanced lithography program distributed an estimated $1 billion to firms exploring next-generation lithography (NGL) approaches.[41] DARPA supported the semiconductor industry's exploration of electron beam lithography, ion projection lithography, and X-ray proximity lithography, as well as extreme ultraviolet (EUV) lithography. EUV ultimately emerged as the preferred NGL technology, with Intel,

---

[38] See, for example, M.M. Waldrop, 2001, *Dream Machine: J.C.R. Licklider and the Revolution That Made Computing Personal*, New York: Penguin Books, p. 201.

[39] See R. Reitmeyer, 1981, "CAD for Military Systems, an Essential Link to LSI, VLSI and VHSIC Technology," *IEEE DAC*, Design Automation Conference, https://ieeexplore.ieee.org/xpl/conhome/10567/proceeding.

[40] G.R. Fong, 1986, "The Potential for Industrial Policy, Lessons from the Very High Speed Integrated Circuit Program," *Journal of Policy Analysis and Management* 5(2), https://www.academia.edu/5757279/The_potential_for_industrial_policy_Lessons_from_the_very_high_speed_integrated_circuit_program.

[41] *EE Times*, 2005, "DARPA Ends Litho Aid at Critical Juncture for Maskless," https://www.eetimes.com/darpa-ends-litho-aid-at-critical-juncture-for-maskless.

AMD, and Motorola funding a PPP for EUV development with three Department of Energy laboratories.[42]

These are only some of the examples of smaller-scale federal semiconductor investments in PPPs that were successful. But by 2020, the United States again faced a loss in chip technology leadership.

CURRENT STATE OF THE U.S. SEMICONDUCTOR SECTOR AS COMPARED WITH OTHER NATIONS

Today's concerns about the decline in U.S. semiconductor leadership arise against the backdrop of the ever-increasing importance of semiconductors to industrialization, economic development, and global prosperity. Other nations have increased R&D investments in the semiconductor sector, and the motivations for these investments are easily understood. More than 50 percent of the global supply of semiconductor chips produced annually is consumed by China today, as it is the largest assembler of consumer electronics.[43] That is, China's prosperity is dependent on sustained access to semiconductor supply, including from U.S. companies, many of which have higher sales in China than in the United States,[44] and China is committed to strengthening and supporting its semiconductor sector across the entire value chain. As has been noted elsewhere, while the United States once clearly led in total science R&D, China is expected to pass the United States in overall governmental R&D investment across all fields in coming years.[45] Asian nations are pursuing carefully defined technology strategies that include R&D and support of applications and production, while the United States historically has avoided such strategies, except for the defense sector, and episodically, in agriculture, transportation and communications, pharmaceuticals and biotechnology, and aerospace.

Although the United States has offshored much of its electronics production, in 2022 semiconductors were still its largest electronic product export and its

---

[42] Intel, 1997, "Government-Industry Partnership to Develop Advanced Lithography Technology," News Release, September 11, https://www.intel.com/pressroom/archive/releases/1997/CN091197. HTM; Intel, 2001, "Partners Unveil First Extreme Ultraviolet Chip-Making Machine," News Release, April 11, https://www.intel.com/pressroom/archive/releases/2001/20010411tech.htm.

[43] D. Araya, 2024, "Will China Dominate the Global Semiconductor Market?" Center for International Governance Innovation, https://www.cigionline.org/articles/will-china-dominate-the-global-semiconductor-market.

[44] D. Butts and S. Chiang, 2024, "China Remains Crucial for U.S. Chipmakers Amid Rising Tensions Between the World's Top Two Economies," *CNBC*, April 12, https://www.cnbc.com/2024/04/12/china-remains-a-crucial-market-for-us-chipmakers-amid-rising-tensions-.html.

[45] National Science Board, Science and Engineering Indicators, 2020, U.S. Trends and International Comparisons, NSBV 20-3, https://eric.ed.gov/?id=ED615506.

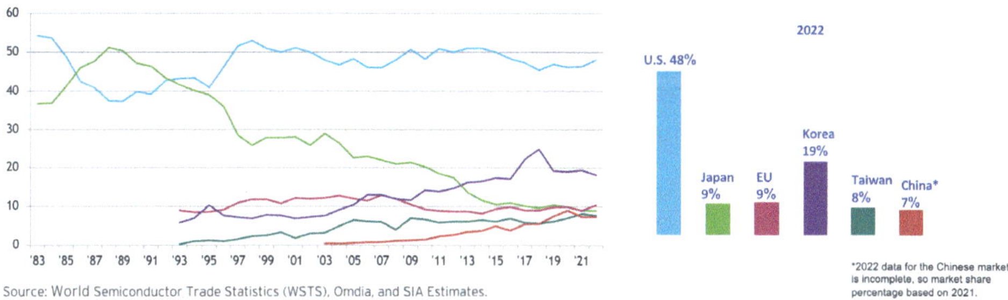

FIGURE 2-2 Global market share of U.S. semiconductor companies, 1983–2022, including from their worldwide facilities.
SOURCE: Courtesy of Semiconductor Industry Association (SIA).

fifth-largest export product, amounting to approximately $61 billion. Companies based in the United States—including both fabless design houses and integrated device manufacturers (IDMs) and their foreign production subsidiaries—held 48 percent of the global semiconductor market in 2022.[46] Figure 2-2 shows that after reversing its decline in global market share in the early 1990s (depicted in Figure 2-1), U.S.-based semiconductor companies continued to capture a stable share (roughly half) of global chip sales from 1996 until today. Semiconductor fabs are located in some 18 U.S. states and employ some 345,000 people.[47]

Despite those impressive-sounding numbers, only 12 percent of the world's semiconductors (by value of sales) were fabricated in the United States in 2022, compared with 37 percent in 1990.[48] Much of the capacity of leading U.S.-headquartered chipmakers is now located offshore. Intel, for example, has 61 percent of its manufacturing capacity offshore, and Micron, the Unites States' only producer of memory chips, has 79 percent of its production capacity overseas.[49]

---

[46] Semiconductor Industry Association (SIA), 2023, *2023 State of the U.S. Semiconductor Industry*, https://www.semiconductors.org/wp-content/uploads/2023/08/SIA_State-of-Industry-Report_2023_Final_080323.pdf.

[47] SIA, 2023, *2023 State of the U.S. Semiconductor Industry*, pp. 9–10; SIA and Oxford Economics, 2021, *Chipping in: The U.S. Semiconductor Industry Workforce and How Federal Incentives Will Increase Domestic Jobs*, SIA, May, https://www.semiconductors.org/wp-content/uploads/2021/05/SIA-Impact_May2021-FINAL-May-19-2021_2.pdf, pp. 11–15.

[48] A. Varas, R. Varadarajian, J. Goodrich, and F. Yinug, 2020, "Turning the Tide for Semiconductor Manufacturing in the US," SIA and Boston Consulting Group, September 6, semiconductors.org/wp-content/uploads/2020/09/Government-Incentives-and-US-Competitiveness-in-Semiconductor-Manufacturing-Sep-2020.pdf.

[49] Hunt, 2022, "Sustaining U.S. Competitiveness in Semiconductor Manufacturing."

Eighty percent of semiconductor sales by U.S. companies are in overseas markets.[50] And most significantly, by 2018, firms in Taiwan and South Korea (i.e., TSMC and Samsung) had taken technology leadership in producing the most advanced semiconductors (at the 5 nm node).[51]

Leadership in Semiconductor Manufacturing by Nations

The estimated share of global production and related efforts carried out in the United States in different semiconductor subsectors, according to a recent Department of Commerce (DOC) study,[52] is as follows:

- R&D: 47 percent
- Design: 27 percent
- Front-end (i.e., wafer-level) fabrication: 12 percent
- Assembly, testing, and packaging: <2 percent, with an additional estimate that 85 percent of chips sold by U.S.-based companies are packaged in Taiwan, China, South Korea, or Malaysia

None of the world's current most advanced chips (at the 5 nm node) were produced in the United States in 2023.

China has invested heavily in semiconductors, starting in the early 2000s.[53] While it does not today have technology leadership, it now has a major presence in older node chip production, and is expected to grow to produce up to 40 percent of world chip supply by 2030.[54] Its government investments specifically in

---

[50] SIA, 2020, "Comments of the SIA on Advanced Notice Regarding the Identification and Review of Controls for Certain Foundational Technologies," November 9, https://www.semiconductors.org/wp-content/uploads/2020/11/SIA-Foundational-Comments-11-9-20.pdf.

[51] The term 5 nanometers (nm) does not refer to actual physical features such as gate length or to transistors being 5 nm in size. Rather, 5 nm is a term used in commercial practice by chipmakers to refer to an improved level of semiconductor chips that reflect a higher degree of miniaturization, increased speed, and reduced power consumption compared to the previous level of 7 nm.

[52] DOC, 2023, *Assessment of the Status of the Microelectronics Industrial Base in the United States*, Bureau of Industry and Security, Office of Technology Evaluation, December, https://www.bis.doc.gov/index.php/documents/technology-evaluation/3402-section-9904-report-final-20231221/file.

[53] See, for example, D. Ernst, 2014, "From Catching Up to Forging Ahead? China's Prospects in Semiconductors," East West Center Working Papers, 1, November, https://www.eastwestcenter.org/publications/catching-forging-ahead-chinas-prospects-in-semiconductors; D. Ernst, 2016, "China's Bold Strategy for Semiconductors," East West Center Working Paper, 9, September 1, https://www.eastwestcenter.org/publications/chinas-bold-strategy-semiconductors-zero-sum-game-or-catalyst-cooperation.

[54] A. Varas, R. Varadarajian, J. Goodrich, and F. Yinug, 2020, "Turning the Tide for Semiconductor Manufacturing in the US," SIA and Boston Consulting Group, September 6, semiconductors.org/wp-content/uploads/2020/09/Government-Incentives-and-US-Competitiveness-in-Semiconductor-Manufacturing-Sep-2020.pdf, p. 10.

semiconductor research and technology development currently dwarf those of the U.S. government.[55] Semiconductors are at the top of China's list of sectors it hopes to dominate in the future—it has already systematically captured major markets in solar, wind, and battery technologies.[56] As China's chip production has ramped up in recent years, there is global concern that it will flood the world market to drive competitors out of the market, then dominate legacy chipmaking (meaning chips with critical dimensions larger than 14 nm), as it has done in other sectors.[57]

The world has become largely dependent on one semiconductor contract manufacturer, Taiwan's TSMC, for producing chip designs that use the most advanced technology available.[58] Today, TSMC produces the majority of the world's high-end logic chips,[59] which are found in virtually all electronic products, from computers to iPhones, and embedded in systems within automobiles and aircraft. With the world's largest fabs, TSMC is the contract manufacturer for many well-known U.S. companies that design their own chips, such as Nvidia and Apple.[60,61] TSMC was formed in 1987 by Morris Chang, a veteran of Texas Instruments, with significant backing from Taiwan's government. TSMC created the dedicated integrated circuit foundry business model, only manufacturing chips for other firms based on those firms' designs. This business model was made possible in part because of the VLSI design systems pioneered for DARPA. TSMC has now

---

[55] P. Triolo, 2021, "The Future of China's Semiconductor Industry," *American Affairs* 5(1), https://americanaffairsjournal.org/2021/02/the-future-of-chinas-semiconductor-industry; S. Ezell, 2021, "Moore's Law Under Attack: The Impact of China's Policies on Global Semiconductor Innovation," Information Technology and Innovation Foundation, February 18, https://itif.org/publications/2021/02/18/moores-law-under-attack-impact-chinas-policies-global-semiconductor.

[56] X. Cen, V. Fos, and W. Jiang, 2022, "Race to Lead: How China's Government Interventions Shape US-China Industrial Competition," Stanford Center on China's Economy and Institutions (paper), August 1, https://sccei.fsi.stanford.edu/china-briefs/race-lead-how-chinas-government-interventions-shape-us-china-industrial-competition.

[57] C. Miller, 2024, "Western Nations Need a Plan for When China Floods the Chip Market," *Financial Times*, January 28, https://www.ft.com/content/2bd1c1a3-931a-4e95-9ea2-e1e8c635ff50.

[58] Y. Jie, S. Yang, and A. Fitch, 2021, "The World Relies on One Chip Maker in Taiwan," *Wall Street Journal*, June 19, https://www.wsj.com/articles/the-world-relies-on-one-chip-maker-in-taiwan-leaving-everyone-vulnerable-11624075400.

[59] Miller, *Chip War*, 197, from Georgetown CSET and Semiconductor Industry Association data.

[60] K. Leswing, 2020, "Apple Is Breaking a 15-Year Partnership with Intel on Its Macs," *CNBC*, November 10, https://www.cnbc.com/2020/11/10/why-apple-is-breaking-a-15-year-partnership-with-intel-on-its-macs-.html.

[61] Intel, which long emphasized production of chips of its own design, moved in 2021 toward a foundry approach, announcing plans for new foundries in Arizona at a cost of $20 billion (S. Nellis, 2021, "Intel to Spend $20 Billion on U.S. Chip Plants as CEO Challenges Asia Dominance," *Reuters*, March 23, https://www.reuters.com/world/asia-pacific/intel-doubles-down-chip-manufacturing-plans-20-billion-new-arizona-sites-2021-03-23).

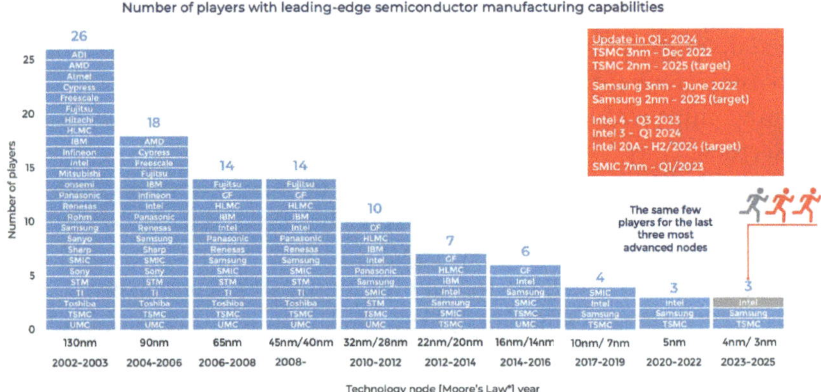

FIGURE 2-3 Semiconductor industry evolution at the leading edge.
SOURCE: Courtesy of Yole Intelligence: *High-End Performance Packaging*: 3d/2.5d Integration (2024).

passed Intel (which did not fully adopt the foundry business model until 2021)[62] as the most important semiconductor manufacturer and has become the world's 11th-largest company. As the first to reach the 7 nm technology node in 2016, and then the first to reach the 5 nm node in 2020, TSMC has demonstrated sustained technology leadership over U.S. chip manufacturers. Today, TSMC makes the majority of the newest cutting-edge logic chips globally, having assumed technology leadership over U.S. chip fabricators.[63] It is also the world's highest volume semiconductor manufacturer, which creates advantages for cost and sustainability.

Due to the large capital costs needed to establish a leading-edge fabrication facility today, only three companies in the world—TSMC, Samsung, and Intel—are still able to compete (see Figure 2-3).

---

[62] Intel, 2021, "Intel CEO Pat Gelsinger Announces IDM 2.0 Strategy for Manufacturing, Innovation and Product Leadership," March 23, https://www.intel.com/content/www/us/en/newsroom/news/idm-manufacturing-innovation-product-leadership.html.

[63] G. Slater and T. Quillin, 2021, "American Technology Leadership Is Critical to Semiconductor Supply Chain Resiliency," *Intel blog*, June 8, https://blogs.intel.com/policy/2021/06/08/american-technology-leadership-is-critical-to-semiconductor-supply-chain-resiliency.

Among these three companies, TSMC leads in making the most advanced chips today, with Samsung and Intel trailing, although threats to this market order exist. Intel has announced its plans to catch up with TSMC. TSMC's location on the island of Taiwan makes it vulnerable to Chinese interference, while Samsung's leading-edge South Korean fabs are within artillery range of a hostile North Korea. Furthermore, China is making significant investments in semiconductor technology in a systematic effort to disrupt the sector.

Other key aspects of the semiconductor supply chain also deserve consideration with regard to economic and national security.[64] China controls a dominant share of certain required critical minerals,[65] and has made major investments in assembly and packaging. Similarly, advanced semiconductor chip test and packaging operations, which enable higher performance, energy efficiency, and density in both leading-edge and some legacy node applications, are provided by firms located almost entirely outside the United States.[66] Compound-semiconductor chips are based on non-silicon substrates largely provided today by firms in Japan, South Korea, and the United States,[67] although Chinese semiconductor firms are working to gain leadership in this area also.[68] These chips are used in power electronics, radio frequency telecommunications, lighting, the electrical grid, electric vehicles, health care devices, and manufacturing technologies today, and have potential to blossom into major new markets.[69]

---

[64] The U.S. role in semiconductor industry sectors is only briefly summarized here. A detailed breakdown of the U.S. share of various global semiconductor sectors is provided in DOC, 2023, *Assessment of the Status of the Microelectronics Industrial Base in the United States*. See also M. Griffith and S. Goguichvili, 2021, "The U.S. Needs a Sustained, Comprehensive and Cohesive Semiconductor National Security Effort," Woodrow Wilson Center, https://www.wilsoncenter.org/blog-post/us-needs-sustained-comprehensive-and-cohesive-semiconductor-national-security-effort, and S. Kahn, A. Mann, and D. Peterson, 2021, "The Semiconductor Supply Chain: Assessing National Competitiveness," CSET, January, https://cset.georgetown.edu/wp-content/uploads/The-Semiconductor-Supply-Chain-Issue-Brief.pdf.

[65] J. Choi, 2023, "US Semiconductor Regulations Spawn a 'Mineral War,'" *Business Korea*, September 8, https://www.businesskorea.co.kr/news/articleView.html?idxno=201318; Z. Yang, 2023, "China Just Fought Back in the Semiconductor Export War," *MIT Technology Review*, July 10, https://www.technologyreview.com/2023/07/10/1076025/china-export-control-semiconductor-material.

[66] IBM has a packaging plant in Bromont, Quebec, Canada; Intel's packaging plants are in Poland and Malaysia. See generally, O. Burkacky, T. Kim, and I. Yeam, 2023, "Advanced Chip Packaging: How Manufacturers Can Play to Win," McKinsey & Company, May 24, https://www.mckinsey.com/industries/semiconductors/our-insights/advanced-chip-packaging-how-manufacturers-can-play-to-win.

[67] Markets and Markets, 2023, "Compound Semiconductor Market," https://www.marketsandmarkets.com/Market-Reports/compound-semiconductor-market-178858112.html.

[68] Tech Insights, 2023, "Can China Dominate the Power Semiconductor Market?" October 16, https://www.techinsights.com/webinar/can-china-dominate-power-semiconductor-market.

[69] See, for example, R.F. Service, 2018, "Beyond Silicon: $1.5 Billion U.S. Program to Spur New Types of Computer Chips," *Science*, July 24, https://www.science.org/content/article/beyond-silicon-15-billion-us-program-aims-spur-new-types-computer-chips.

This shift in production capacity over the past three decades in large part has been owing to financial support provided by foreign governments,[70] which the U.S. government has been unwilling to match until recently. The Semiconductor Industry Association asserts that 70 percent of the higher-total-cost of operations for U.S.-located fabs is driven by industry incentives provided by foreign governments.[71]

As noted above, U.S. firms do not dominate semiconductor manufacturing as they did through the mid-1980s, but they still are highly competitive in many parts of the semiconductor value chain.[72] Although firms in Taiwan and South Korea dominate in advanced logic chip foundry services (TSMC and Samsung), the United States retains strong players in this sector (e.g., Intel, and Abu Dhabi-owned but U.S.-headquartered GlobalFoundries). Likewise, South Korea dominates in memory products (through Samsung and SK Hynix), but U.S.-based Micron remains a strong player in memory products. In chip design, U.S. fabless companies are second to none (these include Nvidia, Apple, Broadcom, AMD, Qualcomm, and many others) and U.S.-based firms Synopsis and Cadence provide world-leading chip design software tools. For semiconductor manufacturing equipment, U.S.-based Applied Materials is a global leader (with about a 20 percent of the global semiconductor equipment market).[73]

Furthermore, where the United States does not lead, the leading companies are often located in friendly countries. For example, Japanese firms have sustained leadership in some key areas, with a global market share of 56 percent in materials (including vitally important photoresists) and 32 percent in manufacturing equipment.[74] Dutch-based Advanced Semiconductor Materials Lithography (ASML)

---

[70] A. Varas, R. Varadarajian, J. Goodrich, and F. Yinug, 2020, "Turning the Tide for Semiconductor Manufacturing in the US," SIA and Boston Consulting Group, September 6, semiconductors.org/wp-content/uploads/2020/09/Government-Incentives-and-US-Competitiveness-in-Semiconductor-Manufacturing-Sep-2020.pdf, p. 17.

[71] SIA and Boston Consulting Group, "Strengthening the Global Semiconductor Supply Chain in an Uncertain Era," Exhibit 16, 34, https://www.semiconductors.org/wp-content/uploads/2021/05/BCG-x-SIA-Strengthening-the-Global-Semiconductor-Value-Chain-April-2021_1.pdf. In Figure 2.3, TCO includes capital expenditure (upfront land, construction, and equipment) plus 10 years of operating expenses (labor, utilities, materials, taxes); "Asian Other" refers to Taiwan and South Korea.

[72] F. Yinug, 2015, "Made in America: The Facts About Semiconductor Manufacturing, Semiconductor Industry Association," SIA, August 2, https://www.semiconductors.org/wp-content/uploads/2020/04/SIA-White-Paper-Made-in-America.pdf.

[73] R. Castellano, 2024, "Applied Materials WFE Market Share Plummets 5% Below ASML in 2023," Seeking Alpha, March 4, https://seekingalpha.com/article/4675800-applied-materials-wfe-market-share-plummets-5-percent-below-asml-in-2023.

[74] N. Tochibayashi and N. Kutty, 2023, "How Japan's Semiconductor Industry Is Leaping into the Future," World Economic Forum, November 20, https://www.weforum.org/agenda/2023/11/how-japan-s-semiconductor-industry-is-leaping-into-the-future.

effectively has a monopoly on the most advanced lithography systems.[75] The CHIPS and Science Act of 2022 (the CHIPS Act) is aimed at strengthening U.S. semiconductor manufacturing, and in sectors of the value chain where the United States is not in a dominant position, there is an opportunity to work to have multiple non-U.S. suppliers in friendly countries.

Additional Challenges: Start-Ups, University Research and Development, and Workforce

Start-ups play a significant role in the U.S. economy generally, but new start-ups in semiconductor fabrication are rare.[76] While entry into advanced-node semiconductor production is inevitably limited because of the very large capital requirements, there are new technology areas where innovation may present opportunities for new entrants, such as a possible boom of chip production to fuel an anticipated expansion of AI applications.[77] Yet only a small portion of total U.S. venture capital has been invested in U.S. start-ups in semiconductor fields,[78] and this has been limited primarily to the seed stage. At the scale-up stage, when start-ups need to bring technologies into production and implementation, foreign capital has been more inclined to invest. For example, among U.S. firms in the semiconductor field seeking to raise Series B- and C-stage capital, a review of 2015 through 2017 shows they had to rely on foreign sources 66 percent of the time.[79] U.S. advances, then, are increasingly controlled by foreign investors, and are easily shifted abroad, leading to what some have called the "innovate here, produce there" story.[80]

In the academic R&D area, while there are strong university departments in semiconductor fields, their ability to conduct research in the most advanced areas is limited by a lack of advanced equipment (an exception is at the SUNY Nanotech campus in Albany, New York).[81] Academic departments simply cannot

---

[75] Miller, *Chip War*, pp. 183–189, 199. See also E. Rasmussen, B. Wilthan, and B. Simonds, 2023, *Report from the Extreme Ultraviolet Working Group Meeting: Current State, Needs and Path Forward*, NIST, Report No. 1500-208, August 22, https://doi.org/10.6028/NIST.SP.1500-208.

[76] President's Council of Advisors on Science and Technology (PCAST), 2022, *Report to the President on Revitalizing the Semiconductor Ecosystem*, White House, September, https://www.whitehouse. gov/wp-content/uploads/2022/09/PCAST_Semiconductors-Report_Sep2022.pdf.

[77] K. Hagey and A. Fitch, 2024, "Sam Altman Seeks Trillions of Dollars to Reshape Business of Chips and AI," *Wall Street Journal*, February 8, https://www.wsj.com/tech/ai/sam-altman-seeks-trillions-of-dollars-to-reshape-business-of-chips-and-ai-89ab3db0.

[78] PCAST, 2022, *Report to the President on Revitalizing the Semiconductor Ecosystem*, pp. 20–23.

[79] Griffith and Goguichvili, Wilson Center.

[80] Ibid.

[81] J.A. del Alamo, D.A. Antoniadis, R.G. Atkins, M.A. Baldo, et al., 2021, "Reasserting U.S. Leadership in Microelectronics—A White Paper on the Role of Universities," Massachusetts Institute of Technology, https://dspace.mit.edu/handle/1721.1/139740.

afford the extremely expensive equipment, so university research, which is critical for advances in many technology fields, is often unable to keep up with industry advances and to adequately contribute to them. This also affects the numbers of semiconductor scientists and engineers that universities can train. Effective arrangements for shared access to advanced facilities between universities and industry have not been formed, so there is a growing disconnect between industry and academic researchers.[82]

Workforce is another challenge.[83] If the United States is seeking to restore its leadership in semiconductor technology, where is the needed talent going to come from? A recent report from the National Security Commission on Artificial Intelligence articulates the solution succinctly: "Cultivating more potential talent at home and recruiting and retaining more existing talent from foreign countries are the only two options to sustain the U.S. lead."[84] Yet while U.S. college enrollment for computer science majors has soared in recent years, enrollment for engineers with the skills needed to work in semiconductor production fields has declined.[85] Salaries in high-paying semiconductor engineering jobs lag behind salaries in software-oriented jobs, where computer scientists and engineers earn more.[86] Computer science and engineering programs continue to grow but are software-focused, and semiconductor hardware design and production is generally not included in the curriculum.[87] Programs to address this imbalance are required if the United States is to support its leading-edge semiconductor manufacturing capacity.[88] Barriers to entry of international talent because of the constraints in the immigration system also affects the available workforce. Equally important is the skilled technical

---

[82] Creating a platform for "chiplet" (an integrated circuit composed of smaller chips where common portions have been designed to support customizable components) design, with supporting software, could focus on designs to demonstrate new innovations and could be one mechanism to improve university as well as start-up design capability. PCAST, 2022, *Report to the President on Revitalizing the Semiconductor Ecosystem*, p. 22.

[83] SIA and Oxford Economics, 2023, *Chipping Away*; NIST, 2023, "R&D Workforce Working Group Update and Recommendations," NIST Industrial Advisory Committee, February 7, https://www.nist.gov/system/files/documents/2023/02/08/Feb%207%20IAC%20Meeting%20Workforce%20Presentation%20Final.pdf.

[84] National Security Commission on Artificial Intelligence, 2021, *Final Report*, https://cybercemetery.unt.edu/nscai/20211005220330/https://www.nscai.gov, p. 173.

[85] D. Martin, 2022, "America's Chip Land Has Another Potential Shortage: Electronics Engineers," *The Register*, July 8, https://www.theregister.com/2022/07/08/semiconductor_engineer_shortage.

[86] J. del Alamo, 2022, "Reasserting U.S. Leadership in Microelectronics," presentation at MIT, May 6, pp. 12–16. See also MIT White Paper, 2021, "Reasserting U.S. Leadership in Microelectronics—The Role of Universities."

[87] B. Nikolic, 2023, "Thoughts on Public–Private Partnerships to Strengthen Semiconductor Design," Presentation to the committee, August 15.

[88] See recommendations in PCAST, 2022, *Report to the President on Revitalizing the Semiconductor Ecosystem*, pp. 16–20.

workforce for semiconductor manufacturing, but there is no comprehensive system in the United States for training and upskilling this workforce.[89] These workforce issues are addressed in more detail in Chapter 7.

INTERNATIONAL COMPETITION: THE ROLE OF NATIONAL SUBSIDIES IN SEMICONDUCTOR FABRICATION

Industrial subsidies by foreign nations, summarized below, have reduced the cost of offshore fabs by some 25 to 35 percent compared to those in the United States. Such industrial support was not seriously considered by the United States for the two decades leading up to the 2022 CHIPS Act. Absent this financing support, advanced fab construction progressively shifted out of the United States, creating a critical issue for economic and national security implications.

In recent years, recognizing the importance of chips for its own economic well-being and national security, China has made very significant government investments in semiconductor manufacturing and R&D.[90] China's firms made 35 percent of non-leading-edge chips, known as legacy chips,[91] in 2019, and that output has been steadily growing, including through sales to the United States. China has been clear on its strategy, which is to first dominate the market for legacy chips, which are still in widespread use in many industrial sectors, while steadily building up expertise and manufacturing capacity as it works toward hoped-for leadership at the most advanced nodes, thereby moving its semiconductor sector further up the value chain. Chinese chipmakers today are

---

[89] S. Shivakumar, C. Wessner, and T. Howell, 2022, "Reshoring Semiconductor Manufacturing: Addressing the Workforce Challenge," Center for Strategic and International Studies (CSIS), October 6, https://www.csis.org/analysis/reshoring-semiconductor-manufacturing-addressing-workforce-challenge; J. Gluck, K. Sodhi, and A. Higuchi, 2023, "Community Colleges and the Semiconductor Workforce," Harvard Belfer Center, June, https://www.belfercenter.org/sites/default/files/files/publication/Community%20Colleges%20and%20the%20Semiconductor%20Workforce_CHIPS%20Series.pdf. See, generally, W.B. Bonvillian and S.E. Sarma, 2021, *Workforce Education—A New Roadmap*, Cambridge, MA: MIT Press.

[90] D. Strub, 2019, "China's Innovation Policy and the Quest for Semiconductor Autonomy—Q&A with Dieter Ernst," *AmCham Shanghai*, May 23, https://www.amcham-shanghai.org/en/article/semiconductor-dieter-ernst; D. Ernst, 2016, "China's Bold Strategy for Semiconductors—Zero Sum Game or Catalyst for Cooperation," East-West Center Working Paper, Innovation and Economic Growth Series No. 9, September, https://www.eastwestcenter.org/publications/chinas-bold-strategy-semiconductors-zero-sum-game-or-catalyst-cooperation.

[91] Daxue Consulting, 2022, "China's Semiconductor Industry Seeking Self-Sufficiency Amid Tensions," September 28, https://daxueconsulting.com/china-semiconductor-industry; Hunt, 2022, "Sustaining U.S. Competitiveness in Semiconductor Manufacturing," p. 2; See, generally, SIA, 2022, "China's Share of Global Chip Sales Now Surpasses Taiwan and Is Closing in on Europe's and Japan's," July 10, https://www.semiconductors.org/chinas-share-of-global-chip-sales-now-surpasses-taiwan-closing-in-on-europe-and-japan.

known to be making every effort to develop 7 nm node logic chips.[92] As part of this overall strategy, there are currently concerns that China's producers will flood the chip market with cut-rate chips (i.e., priced below the actual cost of manufacture) to disrupt the business of current non-Chinese producers, forcing some out of the market, to grow China's share.

The United States has also become reliant on manufacturing facilities abroad for computer memory chips known as dynamic random-access memory (DRAM) and flash memory. Leading-edge DRAM and flash memory fabrication is dominated by fabs located in South Korea, Japan, and Taiwan,[93] although a new U.S. DRAM fab has been announced by Micron.[94] While Chinese companies do not today have leading-edge DRAM fabs, they are already approaching near-leading-edge in flash technology. Flash prices have been falling in recent years, driven by cost reductions as well as by stronger competition than for DRAM.[95] The dynamics of supply chain resiliency and source diversification for DRAM and flash memory is as much of a concern for the United States as for logic chips.[96]

Different nations have used different approaches to support and grow their semiconductor industries. Excepting the case of China, the largest form of government support in semiconductor-producing nations has been tax code measures, typically through R&D tax credits, property tax abatements, offsets to corporate taxes, and investment tax credits. To briefly summarize some of the major support elements, semiconductor firms in Taiwan have benefited from government-supported R&D through its Industrial Technology Research Institute (ITRI) collaborative, budgetary support through the tax code, state-owned shares in semiconductor firms, discounted land and infrastructure support, preferential loans,

---

[92] A. Shilov, 2023, "Huawei's New Mystery 7nm Chip from Chinese Fab Defies US Sanctions," *Tom's Hardware*, August 3, https://www.tomshardware.com/news/huaweis-new-mystery-7nm-chip-from-chinese-fab-defies-us-sanctions; D. Wu and J. Leonard, 2022, "China's Top Chipmaker Achieves Breakthrough Despite U.S. Curbs," *Bloomberg*, July 21, https://www.bloomberg.com/news/articles/2022-07-21/china-s-top-chipmaker-makes-big-tech-advances-despite-us-curbs; Tech Insights, 2022, "7nm SMIC Bitcoin Miner," July 20, https://www.techinsights.com/blog/disruptive-technology-7nm-smic-minerva-bitcoin-miner; P. Alcorn, 2021, "China's SMIC Shipping 7nm Chips Reportedly Copied from TSMC," *Tom's Hardware*, July 21, https://www.tomshardware.com/news/china-chipmaker-smics-7nm-process-is-reportedly-copied-from-TSM-tech.

[93] Hunt, 2022, "Sustaining U.S. Competitiveness in Semiconductor Manufacturing," CSET, pp. 2, 14–15.

[94] Micron, 2022, "Micron Announces Historic Investment of Up to $100 Billion to Build a Megafactory in New York," October 4, https://investors.micron.com/news-releases/news-release-details/micron-announces-historic-investment-100-billion-build-megafab.

[95] See, generally, Imarc, "Flash Memory Card Market Report, 2023–2028," https://www.imarcgroup.com/flash-memory-card-market.

[96] Hunt, "Sustaining U.S. Competitiveness in Semiconductor Manufacturing," CSET, pp. 4, 17–18.

workforce training, and hiring credits.[97] In South Korea, firms benefit from government R&D support through its Electronic and Telecommunications Research Institute (ETRI), budgetary support through the tax code, low-cost financing, infrastructure support, equipment incentives, and workforce training.[98] In China, non-market interventions are much more pervasive. The government has provided semiconductor R&D support and direct funding support of companies through a variety of means, including through state ownership, below-cost financing, below-market equity investment, provision of land at below-market prices, and provision of semiconductor manufacturing equipment.[99] While Europe has fallen behind in advanced chip production, it remains home to one of the most important semiconductor R&D facilities, IMEC, headquartered in Belgium, which has become the leading global center for applied industry development and receives a sustained level of governmental support, which today amounts to approximately 18 percent of its annual revenues. Notably, the United States has no comparable institution at this time. Prior to the CHIPS Act, budgetary support through the tax code was the primary government support vehicle. Historically, U.S. government–funded R&D also played a significant role, such as the investments in SEMATECH until 1995. In something of a return to a direct R&D investment strategy, DARPA has recently invested several billion dollars in its Electronics Resurgence Initiative (ERI), which launched in 2015.[100]

A 2023 study from the Congressional Research Service provides further details on the competitive tools used by other nations in semiconductors; some of its findings regarding Taiwan, South Korea, Japan, Europe, and China are summarized below:[101]

- *Taiwan*. Taiwanese firms dominate leading-edge semiconductor production and play a large role in the global semiconductor supply chain, providing chip design, R&D, semiconductor materials, and assembly, packaging, and testing. Exports of semiconductors amount to some 25 percent of

---

[97] OECD, 2019, "Measuring Distortions in International Markets: The Semiconductor Value Chain," OECD Trade Policy Papers, No. 234, Paris: OECD Publishing, https://read.oecd-ilibrary.org/trade/measuring-distortions-in-international-markets_8fe4491d-en, pp. 17, 60, 47, 54; SIA, 2020, "U.S. Needs Greater Semiconductor Manufacturing Incentives," Chart on Semiconductor Manufacturing Incentives by Country (300 nm or below), July, https://www.semiconductors.org/wp-content/uploads/2020/07/U.S.-Needs-Greater-Semiconductor-Manufacturing-Incentives-Infographic1.pdf, p. 2.

[98] OECD, 2019, "Measuring Distortions in International Markets," pp. 17, 60, 68; SIA, 2020, "U.S. Needs Greater Semiconductor Manufacturing Incentives," Chart on Semiconductor Manufacturing Incentives by Country (300 nm or below), p. 2.

[99] OECD, 2019, "Measuring Distortions in International Markets," pp. 7, 47–54, 68–73, 81, 85–87.

[100] OECD, 2019, "Measuring Distortions in International Markets," pp. 31, 60.

[101] K. Sutter, M. Singh, and J. Sargent, 2023, "Semiconductors and the CHIPS Act: The Global Context," Congressional Research Service, September 28, https://sgp.fas.org/crs/row/R47558.pdf.

Taiwan's Gross Domestic Product (GDP). The government of Taiwan offers incentives to maintain its semiconductor production, including R&D subsidies, tax and tariff stimulus, science parks, subsidized factory buildings, grants and subsidized credit, and connections with local universities and institutes. Examples of tax incentives in 2023 are tax deductions equal to 25 percent of R&D expenditures and 5 percent of spending on new equipment for firms that domestically innovate in critical technologies sectors, including semiconductors. Other recent government actions have included merging two laboratories into the Taiwan Semiconductor Research Institute and providing support for a new ASML lithography tool manufacturing plant.

- *South Korea.* Semiconductors are a critical part of South Korea's production economy, representing 20 percent of its total exports. South Korea has long provided government support, primarily through ETRI (Electronics and Telecommunications Research Institute), a PPP for domestic semiconductor manufacturing. Included in the South Korean strategy is the passage of laws, like the Special Act to Protect and Foster the National High-Tech Strategic Industry (2022), that afford companies regulatory exemptions and tax incentives to spur R&D and increase production output. In 2023, the tax deduction rate was expanded for critical industries, including semiconductors. This included large firms earning tax credits of up to 15 percent and small to midsize firms earning 16 to 25 percent, with the potential to expand by an additional 10 percent. Additionally, South Korea's vertically integrated conglomerates, called chaebols, have facilitated scale-up in market segments, such as DRAMs. The South Korean government has a stated goal of increasing its domestic sourcing of semiconductor manufacturing materials, components, and equipment from 30 to 50 percent by the year 2030. South Korea is also investing directly, negotiating with ASML to build a plant in South Korea. U.S. investments are part of the strategy of South Korean companies: South Korea's Samsung is investing $17 billion in a fabrication plant in Texas and SK Hynix in an $11 billion chip packaging plant.
- *Japan.* Japan maintains a 9 percent market share of the world's semiconductor production. The government has deployed investment initiatives like the "Strategy for Semiconductor and Digital Industries" (2021) aimed at sustaining Japan's current global semiconductor market share and developing new materials beyond silicon, with $6.8 billion in support of domestic semiconductor manufacturing. Japanese firms remain competitive in the production of memory chips, sensors, and power semiconductors. This is evidenced by a 35 percent global market share in semiconductor manufacturing equipment and a 50 percent global market share in the production of semiconductor wafers and photoresists.

- *Europe.* Led by STMicroelectronics and NXP Semiconductors, European-headquartered firms have 10 percent of the global semiconductor market share and have specialized in particular application markets, including the automotive industry, energy applications, and industrial automation. High-value manufacturing equipment leaders like ASML in lithography (Netherlands) and Aixtron in chemical vapor deposition (Germany) help maintain Europe's position in advanced and specialized semiconductor manufacturing equipment. Stable government support for four decades has been vital for the success of IMEC (Belgium), the global leader in semiconductor R&D. Following an $11.3 billion investment in 2013, the European Chips Act, approved in 2023, seeks to mobilize nearly $47 billion in public and private investment, with a goal of Europe having a 20 percent share of the global semiconductor market. In response to this initiative, Intel announced it would invest more than $40 billion for a foundry in Germany, an expanded facility in Ireland, an assembly and packaging facility in Italy, and a new assembly and test facility in Poland, as well as build a major R&D center and make longer-term investments. With support from the French government, in 2022, STMicroelectronics and GlobalFoundries announced plans for a $5.7 billion wafer production facility in France. Between 2023 and 2025, the United Kingdom's government plans to invest up to $245 million, and then more than $1 billion over the following 10 years to support the semiconductor sector in the United Kingdom, with a focus on workforce development to improve the UK industry's access to semiconductor prototyping and tools, as well as strengthening areas of existing capabilities.
- *China.* Many of China's industrial policy approaches toward semiconductors are noted earlier in this chapter. To briefly summarize, those practices include direct industrial production subsidies, grants of land for semiconductor sites, equity support (through government-seeded "guidance funds") and limited regulatory controls; acquiring technology through foreign company acquisitions and joint ventures; setting goals and targets for expansion of both leading-edge and legacy chip semiconductor production; and loose intellectual property practices. A distinguishing feature of China's industrial policy is its heavy reliance on below-market subsidies to capital flowing to Chinese semiconductor companies, in contrast to the directly budgeted tax and subsidy expenditures (e.g., R&D, capital, and other subsidies from both home and foreign governments) received by firms headquartered in other industrialized economies.[102]

---

[102] See Figure 3.2, p. 62 in OECD, 2019, "Measuring Distortions in International Markets."

The most significant reason semiconductor fabrication, assembly, and testing have shifted out of the United States, historically, is because of the manufacturing cost difference between the United States and other nations. The bulk of the cost difference is not in labor costs, which are not a major component of semiconductor fabrication total costs. Rather, studies suggest that differences in costs are predominantly the result of differences in government subsidies (including tax treatment) for semiconductor fabrication facility investments.[103] Prior to the CHIPS Act, U.S. semiconductor producers projected that Asian locations would account for nearly all of expected semiconductor fabrication facility growth.[104] These projections made clear that a next generation of fabs was not going to be built in the United States absent a major, sustained policy and investment intervention.

A THIRD GOVERNMENTAL PUBLIC–PRIVATE PARTNERSHIP IN SEMICONDUCTORS: THE CHIPS ACT

Acknowledging these developments and their security and geopolitical implications, in 2020 bipartisan leaders on the U.S. Senate Intelligence Committee introduced major legislation designed to restore U.S. semiconductor technology leadership. The CHIPS Act, which authorized $52 billion for a series of interventions into the semiconductor sector,[105] was authorized by Congress in 2020 with strong bipartisan support, with appropriations passed in 2022 to fully fund it. This was structured as a one-time bill, with no assurance of follow-on support for ongoing sector needs when the funding ends after 5 years.

The legislation provided $39 billion of financial assistance for companies to construct or modernize U.S.-located semiconductor fabrication plants and to acquire fabrication equipment (see Appendix D). It also offered a 25 percent investment tax credit for investments in U.S.-located semiconductor manufacturing plants and equipment and a $75 billion loan guarantee program. It further provided $11 billion for semiconductor R&D, including funding to set up an industry–university semiconductor technology center, an advanced packaging R&D program, and new advanced manufacturing institutes for semiconductors. The lead agency for all these efforts was DOC. Funding was also provided for DARPA's

---

[103] SIA, "U.S. Needs Greater Semiconductor Manufacturing Incentives," Chart on Semiconductor Manufacturing Incentives by Country (300 nm or below), p. 2; del Alamo, et al., "Reasserting U.S. Leadership in Microelectronics," pp. 17–18.

[104] SIA, "U.S. Needs Greater Semiconductor Manufacturing Incentives," p. 1.

[105] CHIPS for America Act, S.3933, 116th Cong. 2nd Sess., introduced June 10, 2020, https://www.congress.gov/bill/116th-congress/senate-bill/3933. See also Senator Mark Warner, 2020, "Bipartisan, Bicameral Bill Will Help Bring Production of Semiconductors Back to U.S.," Press release, June 10, https://www.warner.senate.gov/public/index.cfm/2020/6/bipartisan-bicameral-bill-will-help-bring-production-of-semiconductors-critical-to-national-security-back-to-u-s.

microelectronics R&D initiatives and for a semiconductor workforce education program at NSF. In addition, funding was available for international supply chain collaborations. Overall, interagency coordination is assigned to the White House Office of Science and Technology Policy's National Science and Technology Council semiconductor subcommittee.

While DOC is the lead agency under the legislation, managing the great bulk of the funding, DoD was not ignored. Prior to the legislation and in many ways anticipating it, DARPA in 2017 began the Electronics Resurgence Initiative (ERI) to develop new semiconductor materials, designs, and architectures;[106] this was renewed in 2023 as ERI 2.0 to promote U.S. technology leadership in next-generation microelectronics research, development, and manufacturing,[107] and DARPA received additional funding in the CHIPS Act for this program. In addition, DoD's Office of the Under Secretary of Defense for Research and Engineering received $2 billion in funding over 5 years under the CHIPS Act to create the Microelectronics Commons (ME Commons) program to create a network of industry–university–government PPPs across the country to focus on technology readiness level (TRL) 4–7[108] semiconductor sector advances. The initial competitive awards totaling $238 million for eight hubs in DoD focus areas were announced in September 2023.[109]

Not long after enactment of the CHIPS Act, three major U.S.-based semiconductor companies—Micron, Qualcomm, and GlobalFoundries—announced large commitments to expand their domestic manufacturing facilities to take advantage of the act.[110] Intel announced major new semiconductor manufacturing fabs in Ohio and Arizona.[111] Samsung (which has also embraced the foundry model)

---

[106] DARPA, 2017, "Beyond Scaling: An Electronics Resurgence Initiative," June 1, https://www.darpa.mil/news-events/2017-06-01.

[107] DARPA, 2023, Electronics Resurgence Initiative 2.0, March 6, https://www.darpa.mil/work-with-us/electronics-resurgence-initiative.

[108] Technology readiness levels (TRLs) are a type of measurement system used to assess the maturity level of a particular technology. Each technology project is evaluated against the parameters for each technology level and is then assigned a TRL rating based on the projects progress. There are nine technology readiness levels. TRL 1 is the lowest and TRL 9 is the highest. https://www.nasa.gov/directorates/somd/space-communications-navigation-program/technology-readiness-levels.

[109] DoD, 2023, "Deputy Secretary of Defense Kathleen Hicks Announces $238M CHIPS and Science Act Award," September 20, https://www.defense.gov/News/Releases/Release/Article/3531768/deputy-secretary-of-defense-kathleen-hicks-announces-238m-chips-and-science-act.

[110] Micron, 2022, "Micron Announces $40 Billion Investment in Leading-Edge Memory Manufacturing in the US," August 9, https://investors.micron.com/news-releases/news-release-details/micron-announces-40-billion-investment-leading-edge-memory; GlobalFoundries. 2022, "GlobalFoundries and Qualcomm Announce Extension of Long-Term Agreement to Secure U.S. Supply Through 2028," August 8, https://gf.com/gf-press-release/globalfoundries-and-qualcomm-announce-extension-of-long-term-agreement-to-secure-u-s-supply-through-2028.

[111] Intel, 2022, "Implementing a Great Step Forward in Competition Policy," August 8, https://community.intel.com/t5/Blogs/Intel/Policy-Intel/Implementing-a-Great-Step-Forward-in-Competition-Policy/post/1406902.

has an existing fab in Austin, Texas, and is planning to open a new one nearby in 2024.[112] TSMC has announced three new fabs in Phoenix, Arizona (its first in the United States), to produce at the 5 and 3 nm scale, although opening has been delayed somewhat owing to workforce recruiting issues.[113]

The CHIPS Act provided a strong signal of U.S unity on the policy goal of retaining a robust semiconductor manufacturing industry deploying state-of-the-art (SOTA) technology. However, with the cost of a single advanced volume-scale fab approaching $20 billion, and roughly equivalent levels of resources required to develop the next generation of fabrication process technology, the level of funding and 5-year commitment in the CHIPS Act will not be sufficient for maintaining an edge capability. Some observers have argued that U.S. manufacturing industry in general has faced hollowing out from decades of disconnect between production and R&D systems; financial sector focus on core competency, which reduced supply chain resiliency; financialization; production offshoring; and underinvestment in advanced production.[114] A number of these problems have seeped into semiconductor production, and it is not clear whether the CHIPS Act will be adequate to reverse these trends. It may stem the decline in chip production facilities in the United States and boost U.S. global production share only modestly. Nonetheless, the act marked a major government intervention into the sector at an unprecedented funding level, even though a longer-term continuing commitment will likely be required to meet the stated objectives.

ACCOMPANYING EXPORT CONTROLS

On October 7, 2022, DOC announced significant export controls on China's semiconductor sector, applying restrictions particularly to exports of tools and equipment used to manufacture semiconductors.[115] The move marked a change in U.S. policy which, starting with China's entry into the World Trade Organization in 2001, had sought to integrate China's economy into a global system on the

---

[112] S. Kim and I. King, 2022, "Samsung Woos US Chip Buyers with Tech Advances, Texas Focus," *Bloomberg*, October 3, https://www.bloomberg.com/news/articles/2022-10-03/samsung-woos-us-chip-buyers-with-tech-advances-texas-production.

[113] H. Charlton, 2021, "Apple's Chip Partner TSMC to Begin Mass Production of 5nm Chips at New Arizona Factory in 2024," *MacRumors*, July 15, https://www.macrumors.com/2021/07/15/tsmc-arizona-factory-begin-production-in-2024; K. Stone, 2023, "TSMC Delays Start of Phoenix Chip Factory, Citing Skilled Worker Shortage," *KTAR News*, July 21, https://ktar.com/story/5519118/tsmc-delays-start-of-phoenix-chip-factory-production-until-2025-because-of-skilled-worker-shortage.

[114] D. Adler and W.B. Bonvillian, 2023, "America's Advanced Manufacturing Problem and How to Fix It," *American Affairs* 7(3), https://americanaffairsjournal.org/2023/08/americas-advanced-manufacturing-problem-and-how-to-fix-it.

[115] J. Schneider and I. Zhang, 2022, "New Chip Export Controls and the Sullivan Tech Doctrine," *ChinaTalk*, October 11, https://www.chinatalk.media/p/new-chip-export-controls-explained.

assumption that a freer Chinese economy would lead to a freer political system and increased adherence to international market laws and norms. With China's increasingly authoritarian rule, expanding military power, and threats to Taiwan, that assumption is increasingly called into question. Following the announcement of the new export controls, stock valuation declined in semiconductor firms both in China (since limiting access to advanced chip production equipment may affect their ability to expand) and the United States (because of concerns that access to Chinese markets that many U.S. firms depend on will be limited).[116] Questions have arisen concerning the initial effectiveness of these controls,[117] and some U.S. semiconductor firms are objecting to further controls because of their dependence on maintaining market share in China.[118] China accounts for approximately one-third of the global semiconductor market and more than $50 billion in combined annual revenue for Nvidia, Intel, and Qualcomm.

The new U.S. export controls are seen by many as part of a global technology "decoupling" that began 7 years ago with China's "Made in China 2025" plan to achieve technology leadership in key technology sectors and make its own economy more self-sufficient.[119] Past history suggests the effectiveness of export controls inevitably fades over time, and the only sustainable way to stay ahead in a technology competition is to retain technology development leadership. This includes maintaining significant R&D with dynamic industry participation, working to bring innovations to market first, and maintaining competitive differentiation. Technology leadership likely needs to be the central thrust of the U.S. and DoD policy approaches, while recognizing that the semiconductor industry is now truly a global one. While the United States may be able to slow the rate of advances in China for a time, leading-edge semiconductor fabrication is unlikely to return to the state of asymmetric U.S. dominance observed at the start of the 21st century, regardless of the CHIPS Act incentives and export controls. Collaboration in

---

[116] I. Morris, 2023, "US Chip Exposure to China Grew Even More Last Year," Chart (from SEC data on U.S. chipmaker companies' sales to China, 2017–2022), *Light Reading*, March 24, https://www.lightreading.com/semiconductorsnetwork-platforms/us-chip-exposure-to-china-grew-even-more-last-year/d/d-id/784025.

[117] T. Fist, L. Helm, and J. Schneider, 2023, "Chinese Firms Are Evading Chip Export Controls," *Foreign Policy*, June 21, https://foreignpolicy.com/2023/06/21/china-united-states-semiconductor-chips-sanctions-evasion; D. Patel, A. Ahmad, and M. Xie, 2023, "China AI & Semiconductors Rise: U.S. Sanctions Have Failed," *Semianalysis*, September 12, https://www.semianalysis.com/p/china-ai-and-semiconductors-rise.

[118] T. Mickle, D. McCabe, and A. Swanson, 2023, "How the Big Chip Makers Are Pushing Back on Biden's China Agenda," *New York Times*, October 5, https://www.nytimes.com/2023/10/05/technology/chip-makers-china-lobbying.html.

[119] R. Fordoohar, 2022, "We Must Prepare for the Reality of the Chip Wars," *Financial Times*, October 27, https://www.ft.com/content/ef90d296-627d-4ff9-9983-ff537bdb078b.

technology development with friendly nations will continue to be valuable for the semiconductor sector, as it is in addressing climate change and other global challenges.

NATIONAL SECURITY AND ECONOMIC SECURITY ISSUES

During the later stages of the Cold War, the U.S. government and DoD, led by Defense Secretary Harold Brown and Under Secretary William Perry, articulated what it called an "offset" strategy whereby the United States would offset Soviet advantages in numbers with superiority in technology.[120] In many respects, this was a renewal of a strategy that dated back to the accelerated, early adoption of cutting-edge technology by the U.S. military in the 1950s and 1960s, as previously discussed. Computer chips were crucial to that strategy, enabling the United States to develop precision weapons that led to a decisive military superiority. Russia effectively missed the information technology (IT) revolution, so was not able to compete with these developing military systems. DoD has found that China's growing technology prowess makes it a greater national security challenge.[121] China's application of an industrial policy approach with very significant subsidies, which the United States has not in the past attempted to match,[122] has allowed China to capture key emerging technology sectors, with semiconductor leadership now a top Chinese priority. China is also turning both more insular and more assertive, with the state reassuming strong control of the technological agenda, moving it away from software-based social media firms toward growing government support and funding for hardware technologies.

Although DoD dominated semiconductor markets in the first two decades of the industry's history, it now accounts for only a tiny share, less than 2 percent, of the total. Furthermore, the bulk of DoD's needs today are for legacy chips: only 11 percent of its spending is for chips with critical dimensions under 28 nm, 18 percent is for chips between 28 and 90 nm, and 66 percent for chips between

---

[120] Office of the Secretary of Defense, 2018, "Harold Brown: Offsetting the Soviet Military Challenge, 1977–1981," Historical Office, March 5, https://history.defense.gov/Portals/70/Documents/speaker_series/webHaroldBrownProgram2018.pdf.

[121] DoD, 2022, "Military and Security Developments Involving the People's Republic of China," Annual Report to Congress, https://media.defense.gov/2022/Nov/29/2003122279/-1/-1/1/2022-Military-and-Security-Developments-Involving-The-Peoples-Republic-of-China.pdf.

[122] G. DiPippo, I. Mazzocco, S. Kenedy, and M. Goodman, 2022, "Red Ink: Estimating Chinese Industrial Policy Spending in Comparative Perspective," Center for Strategic and International Studies (CSIS), May 23, https://www.csis.org/analysis/red-ink-estimating-chinese-industrial-policy-spending-comparative-perspective; D. Adler, 2022, "Guiding Finance: China's Strategy for Funding Advanced Technologies," *American Affairs* 6(2), https://americanaffairsjournal.org/2022/05/guiding-finance-chinas-strategy-for-funding-advanced-manufacturing. Regarding semiconductors, see OECD, 2019, "Measuring Distortions in International Markets," pp. 55–85.

90 and 350 nm.[123] DoD utilizes chips in its systems that are typically two generations or more from the cutting-edge. Its custom-designed application-specific integrated circuit (ASIC) chips are fabricated in approved "trusted foundries" in the United States, for security reasons, on manufacturing lines that often lag even further from the leading-edge. These facilities will continue to be important for DoD, especially as innovation at legacy nodes continues—for example, in automotive applications.

However, new defense-related technologies are very likely to push DoD toward increasing reliance on more advanced chips. The military will require use of the most advanced available artificial intelligence (AI) coprocessor chips as AI capabilities are deployed for applications ranging from automating development of software and firmware to being inserted into military systems as a critical new tool of warfare. DoD will also require leading-edge chips for its advanced data fusion centers, autonomous platform technologies, advanced sensor systems, distributed cloud computing, and promising defense applications emerging from quantum IT. These advances will require DoD to field new software and systems, running on chips with the lowest possible power consumption, lowest possible weight, and greatest possible density, to respond to potential adversaries. In addition, because weapons platforms like aircraft and ships must last for decades, over time these must be modernized to incorporate new capabilities, and DoD needs advanced chips for these modernization initiatives. Chiplets and other more advanced electronics capabilities are expected to emerge more rapidly in the future; therefore, software and hardware updates need to be accommodated effectively and efficiently on a much more frequent basis, as well as factored into the design process for new equipment.

Other advances are on the horizon. On-chip silicon structures have been evolving from a two-dimensional (2D) approach to a focus on three-dimensional (3D) architectures. The heterogeneous integration of "chiplets" into 3D packages containing multiple chips, perhaps each produced using different processing technologies and materials, is evolving rapidly.[124] New computer architectures that overcome energy efficiency limits and eliminate barriers to rapid access to data in memory, photonic interconnects, electronic design automation to make custom

---

[123] DOC, 2023, *Assessment of the Status of the Microelectronics Industrial Base in the United States,* Table, Respondent End Use Projections—Percent of Chips Segment Revenue, U.S. Defense; Table, Respondent End Use Projections—Percent of End Use Revenue for Given Nodes, U.S. Defense. See also D.J. Radack, B.S. Cohen, R.F. Leheny, V. Sharma, and M.M.G. Slusarczuk, 2016, *Semiconductor Industrial Base Focus Study—Final Report,* Institute for Defense Analyses.

[124] See, for example, S. Zhang, Z. Li, H. Zhou, S. Wong, R. Li, S. Wang, K. Paik, and P. He, 2022, "Challenges and Recent Prospectives of 3D Heterogeneous Integration," *e-Prime—Advances in Electrical Engineering, Electronics and Energy* 2, https://www.sciencedirect.com/science/article/pii/S2772671122000249.

circuit design less expensive, and specialized architectures that optimize computing performance for specific applications are being actively explored.[125] As current transistor device technologies hit scaling limits over the next decade, new post-complementary metal-oxide-semiconductor (post-CMOS) transistor technologies are emerging.

Inevitably, DoD will increasingly be functioning as a high-end customer in the advanced chip market. DoD grasps this reality; the security of the industrial base on which it depends is increasingly a key focus for national security.[126] That is, today DoD understands that the lack of a healthy advanced semiconductor manufacturing sector in the United States will compromise its defense capability.

It is clear to the committee that the United States will not be able to replicate a standalone national supply chain—entrenched specialization and comparative advantage across the supply chain and steep capital and R&D costs preclude this. Given that semiconductor independence is not economically feasible, DoD needs to determine which strategic semiconductor partnerships it requires to assure a diversified and reasonably secure supply chain, with both redundancy and competitive access. The United States has many strengths, including its R&D base, the remaining U.S. firms, the depth of its market, and its ability to create global alliances guaranteeing security in semiconductor manufacturing capability.

The committee is confident that the United States can retain a significant and robust position in leading-edge, commercial semiconductor manufacturing with new strategies and support mechanisms. Key to success will be ensuring some protection for critical domestic links in supply chains and nurturing key U.S. production elements—that is, "onshoring." However, a secure ecosystem will also require a network of multiple suppliers located in both the United States and in friendly nations to meet DoD objectives, which means maintaining a complex web of multinational relationships—"friend-shoring."[127]

The CHIPS Act has a 5-year focus, whereas the issues described above are longer-term challenges that DoD faces. Therefore, there are two ongoing transitions: a transition to new DoD system designs that incorporate the most advanced semiconductors available, which is paralleled by preparation for more

---

[125] OSTP, 2022, "Draft National Strategy for Microelectronics Research," pp. 8–9.

[126] DoD, 2022, National Defense Strategy, October 27, https://www.defense.gov/National-Defense-Strategy; OSTP, 2022, "Draft National Strategy for Microelectronics Research, Comments of the SIA on OSTP Request for Information," October 17, https://www.semiconductors.org/wp-content/uploads/2022/10/OSTP-National-Micro-Strategy-RFI-Response-SIA-FINAL.pdf, pp. 1–2.

[127] OSTP, 2022, "Draft National Strategy for Microelectronics Research," pp. 1–2, 4. See also SIA and Boston Consulting Group, 2021, "Strengthening the Global Semiconductor Supply Chain in an Uncertain Era," SIA, April, https://www.semiconductors.org/wp-content/uploads/2021/05/BCG-x-SIA-Strengthening-the-Global-Semiconductor-Value-Chain-April-2021_1.pdf, pp. 27–38.

disruptive technologies on the horizon. DoD has in recent decades often chosen to prioritize access to secure chips over access to advanced chips; its need for upcoming disruptive technology advances enabled by advanced semiconductors will require it to rethink this balance, as discussed in Chapter 5. DoD will also need strategic international partnerships to achieve these objectives throughout the coming decade.

LESSONS FOR THE DEPARTMENT OF DEFENSE FROM SEMICONDUCTOR HISTORY

Technological and economic drivers have been central to the evolution of the global semiconductor industry over the last five decades, and understanding how they have changed is critical to a successful DoD semiconductor strategy.

Leading-Edge Chip Costs Are Declining More Slowly Compared with Previous Trends

The primary technological reality from about the early 1960s through the first decade of the 21st century was that the area occupied by an electronic device on a silicon chip (transistor density), the speed of the device (transistor switching time), power used per electronic device (energy used per transistor), and manufacturing cost per device (transistor cost) all improved rapidly, in concert, owing to what came to be known as Dennard scaling.[128] However, the steady flow of technological improvements for creating 2D structures on the surface of a silicon wafer has slowed considerably over the past decade, as leading-edge fabrication technology shifted toward more complex 3D electronic device structures. The net result is that while device size, speed, and density continue to improve with new generations of silicon fabrication technology, cost declines have lessened compared with previous historical trends. Figure 2-4 reflects one large semiconductor company's view of electronic device performance and costs at recent technology nodes.[129]

---

[128] The underlying theory of Dennard scaling suggested that a 30 percent reduction in transistor feature size, associated with a 50 percent reduction in transistor area, would be accompanied by a 30 percent reduction in delay (40 percent increase in clock frequency) and a 50 percent reduction in power. See H.E. Esmaeilzadeh et al., 2013, "Power Challenges May End the Multicore Era," *Communications of the ACM* 56(2):95.

[129] For additional discussion on the flattening out of cost declines at recent nodes, including data for both fabless and foundry semiconductor companies, see K. Flamm, 2018, "Measuring Moore's Law: Evidence from Price, Cost, and Quality Indexes," Working Paper 24553, National Bureau of Economic Research, https://doi.org/10.3386/w24553, April; C. Corrado, J. Haskel, J. Miranda, and D. Sichel, eds., 2021, *Measuring and Accounting for Innovation in the Twenty-First Century*, Chicago: University of Chicago Press and National Bureau of Economic Research, pp. 417–419.

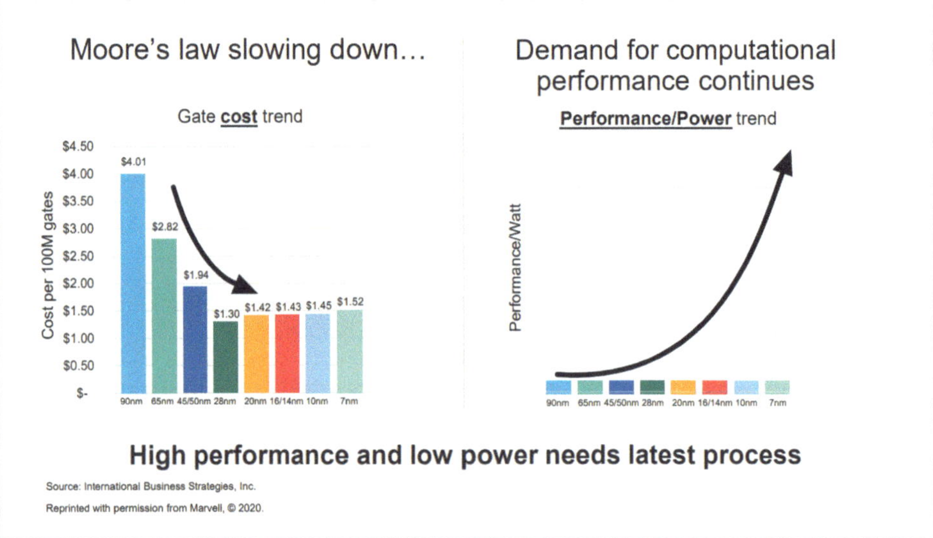

FIGURE 2-4 Cost per 100 million logic gates at recent technology nodes.
SOURCE: Adapted from Marvell Technology, Inc., Investor Day 2020 Presentation, https://filecache.investorroom.com/mr5ir_marvell/131/download/Marvell%20Investor%20Day%202020.pdf. Reprinted with permission from Marvell, © 2020.

For DoD to achieve its goal of uncontested technological superiority over potential adversaries—which has been the central organizing tenet underlying post–World War II national security strategy—it continues to seek weapons systems that achieve the highest possible performance with the lowest possible power requirements and with the highest possible density (smallest size and lightest weight). With the electronics that power this technological advantage no longer falling in cost at historic rates, or perhaps even increasing in cost, DoD should be prepared to once again invest substantial sums in achieving the technological superiority needed to offset numerically superior adversaries, as it did in the 1950s and 1960s.

Economies of Scale

Increasing economies of scale (i.e., the average cost per unit declines with increased output of units) in chip fabrication now make leading-edge chip designs relatively more expensive in low volumes than in the past, primarily because of fixed production costs that do not vary with production volume. That is, costs per chip are now often dominated by very large, fixed costs. The principal fixed cost drivers for increasing economies of scale in chip fabrication are as follows: the ever-greater investments needed to build a leading-edge fab, the escalating costs of creating mask sets

R&D for chips and fab module construction costs are soaring.

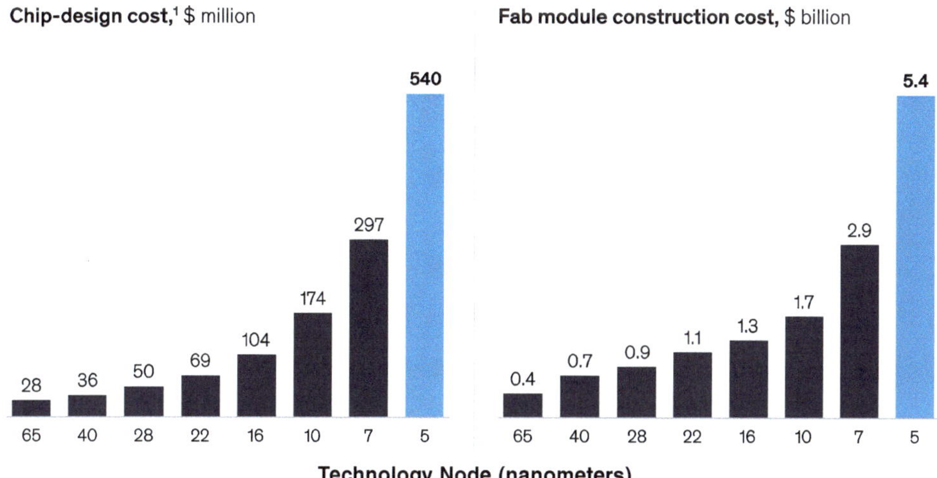

FIGURE 2-5 Increasing fixed costs in chip production as a function of technology node.
NOTES: Footnote 1 defined as major components include intellectual property qualification, architecture, verification, physical, software, prototyping, and validation. R&D, research and development.
SOURCE: Courtesy of McKinsey & Company: Industrials and Electronics.

(reticles), the larger numbers of masks and processing steps required; and the soaring fixed costs of designing, verifying, prototyping, and validating new chip designs. (See Figure 2-5 for design costs per chip and investment costs per wafer fab module.)

The net effect of rapidly increasing fixed investment costs for the newest technology nodes has been to create economic advantage for the largest volume manufacturers, which in turn has led to industry consolidation over time. Today, the per-wafer costs of manufacturing semiconductors are much higher at low volumes of output than historically has been the case (see Figure 2-6). While there are many more transistors per wafer at the most advanced technology node, Figure 2-4 illustrates that cost per leading-edge transistor for even large fabless semiconductor producers like Marvell has risen at the typical volumes at which their most advanced chip designs are manufactured and sold.

Indeed, there has been increasing concentration in virtually every segment of the semiconductor industry (including semiconductor equipment and materials) in the 20th and early 21st century, driven by economies of scale. In the 20th century postwar period, many computer manufacturers had their own captive fabs churning out custom chips for use in their computer products. Increasing economies of scale led those computer companies to outsource their chip designs to foundries and drop out of chip manufacturing one by one—with the last to leave

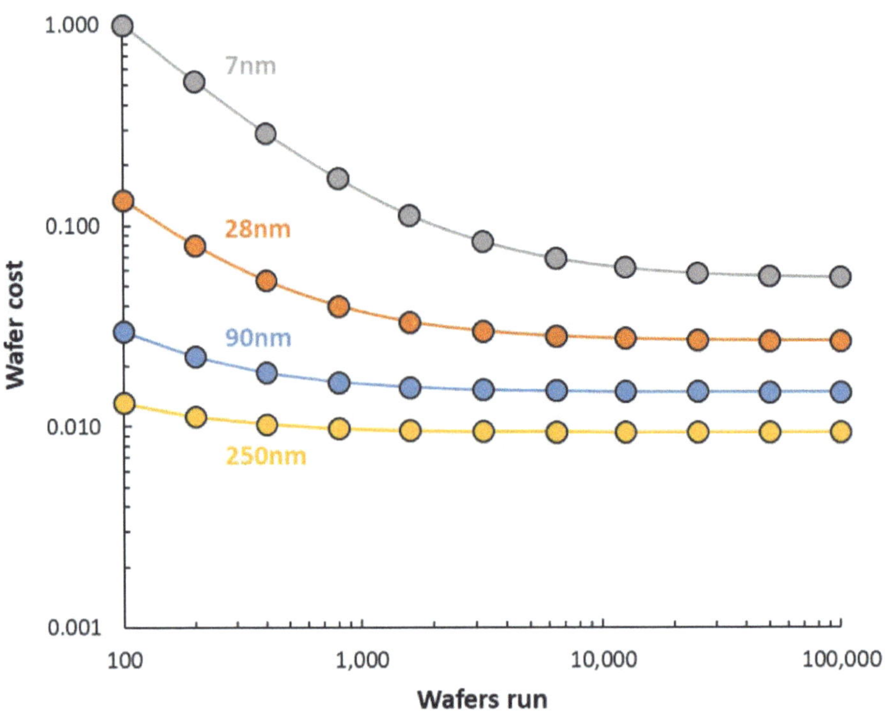

FIGURE 2-6 Economies of scale in chip fabrication have increased at recent technology nodes. SOURCE: Courtesy of Dylan Patel of SemiAnalysis.

being IBM, in the 21st century. Many companies followed the same path, shifting to so-called "fab-lite" models or even "fabless" business models. Most of today's U.S. semiconductor industry sales are composed of semiconductors designed in the United States but manufactured at facilities scattered around the globe, using inputs sourced from a thoroughly globalized materials and equipment industry. These developments create growing cost containment problems for DoD, which is often in the position of being a low-volume customer seeking customized chips.

Investments in Research and Development

Historically, U.S. technology leadership has led to better outcomes than efforts to control the global diffusion of technology through export controls, which are costly in terms of political and economic capital and rarely work over the long term. There are notable examples of this pattern within the semiconductor sector. Development of germanium transistors in the United States was quickly copied abroad in the 1950s. Rather than impeding imports, the U.S. semiconductor industry invested in development of silicon transistors, which ultimately displaced the lower-performance germanium devices. When foreign producers became competitive in silicon transistors, U.S. chip producers moved on to integrated circuits. When foreign producers successfully copied—and improved on—U.S. memory integrated circuits, U.S. producers moved on to new kinds of circuits, including microprocessors and digital signal processors. When foreign computer manufacturers succeeded in producing competitive mainframes and supercomputers, U.S. producers developed new machines using novel parallel architectures, high-performance processor interconnection fabrics, and other architectural innovations, like reduced instruction set computers (RISC).

As noted, a major transition to post-CMOS technologies is now in view. Experience shows that export controls can create a measure of success in slowing down and making more costly the technological development of a rival or potential adversary in the short- to medium-run. But historically, export controls have not prevented the diffusion of technology in the long-run. However, strategic advantage can only be maintained by investing in the next generation of technology. This suggests that DoD's interests will be best served not by relying simply on export controls but also by striving for technology leadership, which creates both strategic advantage for DoD (when inserted into DoD systems) and competitive advantage for the U.S. firms pioneering new products serving both commercial and defense applications.

It is important to recognize that R&D in both next-generation (i.e., post-CMOS, as discussed further in Chapter 3) semiconductors and in design innovation will be important to DoD's continued success. The CHIPS Act focused on manufacturing facilities, but with chip design costs soaring at recent technology nodes, reducing the cost of chip designs for specialized DoD needs is another critical problem for DoD. DoD is not alone; this problem is shared by start-ups and small to midsized firms trying to establish markets for innovative system designs that require advanced chips. This is a key area where strong alignment and better program integration between DoD and DOC can yield substantial economic and security benefits for the nation.

3

Reducing Barriers to Sustainable and Resilient Semiconductor Production

In addition to a review of the competitive position of the U.S. semiconductor sector, the committee was asked to review barriers to sustainable and resilient U.S. semiconductor production, including the factors that drive production and create reliable supply chains for equipment, components, and talent. This topic is discussed in this chapter with a particular focus on the Department of Defense's (DoD's) semiconductor sector needs and its potential role in strengthening this system.

In 2006, DoD's Defense Science Board (DSB)[1] observed that three categories of electronics or semiconductor capabilities are relevant to national security electronics requirements: semiconductors used in applications where chips do not create a strategic competitive advantage; commercial (dual-use) leading-edge semiconductors whose use in military applications is vital to the performance of that equipment; and military-specific (defense-unique) semiconductors that are required to create leading-edge capabilities for defense systems. Only the latter two categories are primarily relevant to DoD concerns about maintaining its qualitative technological superiority relative to potential adversaries.

The previous discussion makes clear that, in contrast to the situation in 2006, the availability of leading-edge, dual-use semiconductors (the second category above) for use in DoD systems can no longer be taken for granted. In 2006, the DSB noted that state-of-the-art (SOTA) semiconductor fabrication plants (fabs)

---

[1] See Chapter 4 in U.S. Defense Science Board (DSB), 2006, *Joint US Defense Science Board-UK Defence Scientific Advisory Council Task Force on Defense Critical Technologies*, March, https://dsb.cto.mil/reports/2000s/ADA446196.pdf.

"are being established in the Far East" and that the United States "no longer has the asymmetric technology advantage that it once had." The report went on to note that the "United States is not behind. But it is no longer as far ahead as it once was. One consequence of this is that the SOTA COTS [commercial-off-the-shelf] technology is available to U.S. adversaries." In contrast to the situation observed in 2006, the committee believes that the United States *is* clearly now behind in leading-edge semiconductor fabrication, and the asymmetric technological advantage in semiconductor manufacturing resides outside the borders of the United States. Intel is striving to catch up on the most advanced chips, but DoD will also need to partner with the international firms that today provide the most advanced leading-edge microelectronics. This will require DoD to balance its desire to procure secure and "trusted" chips with its desire to field systems that have the most advanced chips. In a situation of continuing foreign company leadership, it can have one or the other, it cannot have both in the same chip.

The 2006 DSB task force went on to note that while some worried that the United States "will not have an assured and trusted supply of SOTA integrated circuits (ICs)," the task force's view was that "in the next 10–20 years, wafer manufacturing in the United States will remain sufficiently strong to support DoD needs in the event of supply disruption in the Far East. One key reason is that the cost of wafer fab in the Far East is not substantially cheaper than the cost of wafer fab in the United States. Wafer fab costs are dominated by capital depreciation, and low cost labour [sic] has minimal impact on wafer fab cost."[2]

The DSB task force did not in 2006 foresee the large subsidies to capital investment in semiconductor manufacturing that the U.S. semiconductor industry argues have been deployed by foreign competitor nations. Wafer fab costs, it appears, now *are* cheaper in other nations, due in large part to national industrial policies. The DSB in 2006 recommended that "DoD should also monitor the number of new U.S.-based fabs and provide incentives for U.S. companies to locate fabs in the United States as required."[3] In the year following the release of this report (2007), 16 percent of global semiconductor wafer fab capacity was located in the United States.[4] By 2022, that figure was down to 10 percent,[5] the U.S. share of fabs using more modern equipment, processing 300 mm wafers, was 8 percent,[6] and the U.S.

---

[2] DSB, 2006, "Joint Task Force on Critical Technologies," pp. 67–68.

[3] DSB, 2006, "Joint Task Force on Critical Technologies," p. 70.

[4] National Research Council, 2012, *Rising to the Challenge: U.S. Innovation Policy for the Global Economy*, Washington, DC: The National Academies Press, https://doi.org/10.17226/13386, p. 340.

[5] Semiconductor Industry Association (SIA) and Boston Consulting Group (BCG), 2024, "Emerging Resilience in the Semiconductor Supply Chain," SIA, May, https://www.semiconductors.org/wp-content/uploads/2024/05/Report_Emerging-Resilience-in-the-Semiconductor-Supply-Chain.pdf, p. 14

[6] SEMI 300mm Fab Outlook to 2026, cited at https://www.statista.com/chart/31371/distribution-of-global-semiconductor-fabricating-capacity.

share of fab capacity using leading-edge (less than 10 nm) technology nodes was 0 percent.[7]

The challenge at hand is how to construct a system of incentives that guarantees DoD and U.S. industry continued access to at least one leading-edge, high-volume commercial foundry within the United States over the next 10–20 years, and the continuing development of follow-on advanced chip manufacturing technology onshore beyond the 5-year time horizon of the CHIPS and Science Act of 2022 (CHIPS Act).

As previously discussed, the new generations of semiconductor chips are no longer getting cheaper at the rates they did under the era of Moore's Law scaling paradigm. However, increased transistor miniaturization and lower-energy consumption continue to advance. For DoD, lowering the energy required for computations is now critical for placing greater intelligence in munitions or piloting drones and autonomous systems. In fact, DoD has a clear need to adopt chip architectures that enable lower energy use, greater density, and lower weight for the computing and artificial intelligence it will be embedding in its most advanced systems. Making it easier and cheaper to design and implement new systems and software and realize those designs with the most advanced chip manufacturing technology nodes is a growing DoD priority.

Achieving the outcomes DoD requires means overcoming a series of barriers to those outcomes. The most critical are described below.

OVERCOMING BARRIERS TO MANUFACTURING

Because it requires low volumes, and sometimes customization of chips, DoD is not in a good position to make demands on an industry that mainly relies on producing very large volumes of standardized chips. On the other hand, there is good reason to believe that there are many specialized new commercial chip designs targeting lower-volume markets that have shifted away from utilizing the most advanced available fabrication technology as Moore's Law slowed.[8]

---

[7] SIA and BCG, 2024, "Emerging Resilience in the Semiconductor Supply Chain," p. 14.

[8] As fixed design and mask costs for leading-edge fabrication technology nodes have increased, new low-volume applications have sometimes found mature, older fabrication technology to be the most economic choice. This resulted in a post-2014 "reawakening" wave of fabrication capacity investment using older vintage equipment from more mature technology nodes (e.g., fabrication equipment compatible with smaller 200 mm wafers). See the discussion in K. Flamm, 2021, "Measuring Moore's Law: Evidence from Price, Cost, and Quality Indexes," pp. 447–448 in *Measuring and Accounting for Innovation in the 21st Century* (C. Corrado, J. Miranda, J. Haskel, and D. Sichel, eds.), NBER and University of Chicago, https://www.nber.org/system/files/chapters/c13897/c13897.pdf.

Economic success in advanced chip manufacturing depends on a high-volume/low-mix business model, creating a fundamental challenge for DoD with respect to its needs for at least some custom, leading-edge, low-volume/high-mix, microelectronic components.[9] If DoD-funded initiatives can make designing application-specific integrated circuit (ASIC) chips substantially less expensive using the most advanced fabrication technology nodes, as will be required for insertion of custom leading-edge chip technology into new defense systems, it could also have a substantial impact on the commercial chip industry. The same advances that make it less costly for DoD to design ASICs would also enable low-cost designs of commercial and industrial ASICs that use the most advanced available fabrication technologies.

DoD will need continuing access to leading-edge chips. Because U.S. firms have lost leadership in the most advanced logic chip production,[10] it is important from both an economic and a national security perspective for the U.S. government to continue to encourage the current leaders, TSMC and Samsung, to create advanced manufacturing capabilities in the United States. At the time of writing, Intel is working to produce a new generation of advanced chips,[11] and TSMC and Samsung are now building new fabs in Arizona and Texas, respectively. Both TSMC and Samsung are expected to continue to produce the bulk of their chips in their home countries, but having at least some production capacity for the most advanced chips in the United States is important to DoD future technology needs.[12] Reflecting this, DoD has formed the new Office of Strategic Capital (OSC) to support the scale-up

---

[9] In contrast to chip design costs, to some extent the rapidly increasing fixed costs of the masks (reticles) used to fabricate designs can be shared across different designs by bundling multiple designs together on a single multi-product "shuttle" wafer. The number of chip designs that can be bundled together on a single wafer is limited by the size of the mask (reticle) used to pattern a wafer. EUV reticles are currently limited in size to under 900 mm^2, and reticle size in upcoming new generations of equipment is likely to further decrease. See https://en.wikichip.org/wiki/mask.

[10] M. Deutscher, 2023, "Intel Begins Mass-Producing Chips Using Cutting Edge EUV Technology," *Silicon Angle*, September 29, https://siliconangle.com/2023/09/29/intel-begins-mass-producing-chips-using-cutting-edge-euv-technology; J. Laird, 2023, "Intel's 18A 'Hallelujah' Moment Is the Biggest Bet the Company Has Ever Made," November 13, https://www.pcgamer.com/intels-18a-hallelujah-moment-is-the-biggest-bet-the-company-has-ever-made.

[11] See, for example, M. Ahmad, 2023, "S Adds Two Variants to Its 2nm Node, Will Intel Catch Up?," EDN, May 3, https://www.edn.com/tsmc-adds-two-variants-to-2-nm-node-will-intel-catch-up.

[12] CHIPS Act funding helped encourage this development, assisting in offsetting the 20 to 30 percent cost differential for production in the United States (largely from, as noted above, subsidies that competitor nations' industrial policies provide to their semiconductor sectors). However, because CHIPS Act support expires in 5 years, a longer-term manufacturing perspective is required. This means that governmental financing for this differential likely needs to continue past this horizon. Continuation of at least the investment tax credit and loan guarantee programs in the CHIPS Act must therefore be considered.

and production of critical defense-relevant technologies. As stated in DoD's *Investment Strategy for the Office of Strategic Capital: Fiscal Year 2024*:[13]

> The mission of the Office of Strategic Capital (OSC) is to attract and scale private capital to technologies critical to the national security of the United States. OSC's initial emphasis is on using loans and loan guarantees in partnership with other federal departments and agencies to crowd in capital for component-level technologies.
>
> OSC is unique but complementary to existing DoD efforts through the combination of three primary areas of emphasis. While not an exhaustive list, these areas include:
>
> 1. **Components** (not capabilities)
> 2. **Finance** (not innovation)
> 3. **Lending** (not spending).

Provided it has sufficient resources, OSC could play an important role in the semiconductor ecosystem through provision of much-needed patient capital to bring new technologies to maturity.[14] As stated earlier, a robust partnership between DoD and DOC is of the utmost importance to ensure continued support for U.S. manufacturing past the current CHIPS Act funding horizon.

Extension of Financing to Assure Continuing Advanced Chip Production

The CHIPS Act is a 5-year budget authorization, and there is no assurance that it will be extended, although the committee anticipates that the problems of U.S. chip leadership and the need for governmental financing to assure production capability in the United States will extend well beyond that. Therefore, DoD (with DOC) needs to consider ways to extend financing mechanisms for advanced semiconductor facilities over the longer term, including subsidies, loan guarantees, and investment tax credits. (See Recommendations 5.12 and 5.13.)

In this regard, it is worth recalling that the Defense Advanced Research Projects Agency (DARPA) was involved in creating the original foundry model, the Metal Oxide Semiconductor Implementation Service (MOSIS) for prototyping. DoD invested considerable resources in development of computer-aided design software tools and network infrastructure connecting university and industrial customers to semiconductor fabricators, utilizing design rules based on currently available fabrication technology and manufacturing processes. This DoD-underwritten infrastructure ultimately blossomed into the current commercial foundry ecosystem for both mature and leading-edge technology nodes. The foundry model now dominates the industry.

---

[13] DoD, 2024, *Investment Strategy for the Office of Strategic Capital: Fiscal Year 2024*, https://www.cto.mil/wp-content/uploads/2024/05/OSC_FY24_Investment_Strategy.pdf, p. 1.

[14] DoD, 2022, "Secretary of Defense Establishes Office of Strategic Capital," December 1, https://www.defense.gov/News/Releases/Release/Article/3233377/secretary-of-defense-establishes-office-of-strategic-capital.

As a relatively low-volume customer aspiring to use leading-edge technology in its systems, DoD cannot avoid utilizing the foundry model it was so instrumental in creating in order to address its current semiconductor sourcing needs.

Extensions of the investment tax credit and loan guarantee programs can help assure DoD of continued access to an overall foundry manufacturing base in the United States for SOTA semiconductors. But then another issue arises: DoD also needs to work with *specific* semiconductor foundry companies to assure a supply that meets its particular needs for custom chips, including in emergencies. With few exceptions, the foundries that exist today to primarily meet DoD demand are not at the cutting-edge and therefore are not commercially competitive and lack high-volume, leading-node production.

To be commercially competitive and operating at the leading node, foundries cannot exist to primarily meet DoD needs. This reality requires DoD to forgo some of the security controls that it has historically required in hardware supply chains. Accepting some alternative control systems with additional information security risk that comes with these commercial partnerships is necessary given today's status quo alternative, which limits DoD access to the most advanced chips it requires for defense systems. This challenge is discussed further in the section below on regulatory barriers.

Foundry Characteristics Needed for Department of Defense Access

DoD has several unique needs for U.S.-located foundries manufacturing leading-edge semiconductors, including dual-use production, ability to service high- and low-volume customers, competitive incentives to remain at the leading edge, sensitivity to international supply chains, and managing supply chain risks.

- *Dual use.* Only a manufacturer successfully serving commercial markets will have the volume needed to amortize the escalating cost of leading-edge fabrication facilities and R&D on the next generation of technology. Accordingly, DoD is incentivized to use foundries that are dual-use manufacturers, serving both DoD and commercial customers.
- *High and low volume.* Such a foundry will serve both high-volume and low-volume customers, for commercial and for defense needs. Economical production of low-volume chip designs at the leading edge (including prototype and research chips) is possible today when several designs are aggregated together onto a single, large, multi-project wafer, thus dividing the fixed costs of mask sets and production overhead cost among multiple projects.
- *Maintaining competition.* Ideally, there will be multiple foundry producers located in the United States or its trusted allies capable of serving both defense and commercial customers, competing freely against one another at the leading edge, including for DoD business. (The track record of sole-supplier

"national champions" in providing viable, competitive high-tech products is highly problematic.) Maintaining some significant degree of competition, with at least some of that leading-edge capability in the United States, is worth some additional cost from both the national security and economic security perspective.

- *International supply chains.* Increased economies of scale and the globalization of semiconductor infrastructure mean that geopolitical risks to supply chains can be managed but cannot be eliminated. Semiconductor industry companies headquartered in the United States, as well as companies headquartered in friendly nations, all rely on R&D groups and regional centers of excellence scattered around the globe. IBM's development of metal-oxide-semiconductor (MOS) memory technology in the late 1960s and 1970s made critical use of a German research laboratory; some of Intel's important processor designs were developed in Israel; equipment producer Applied Materials depends on its Finnish subsidiary for specialized expertise in atomic layer deposition and etch.[15] Requiring that all components, expertise, and skills be sourced within national boundaries for national security reasons would be prohibitively expensive and would likely lead to a loss of technology leadership.

- *Community solution.* A coalition of "friends and allies" is an appropriate approach for managing supply chain risks to both national security and economic security.

Partnership Mechanisms for Department of Defense Fabrication Plant Access

DoD's Rapid Assured Microelectronics Prototypes (RAMP) program and, more recently, its follow-on RAMP-C program illustrate approaches that DoD can apply in partnering with manufacturers to meet its chip needs.[16] Starting in 2020, DoD's RAMP program demonstrated how to securely obtain SOTA microelectronics technologies from industry without depending on a closed security

---

[15] See S. Leibson, 2023, "A Brief History of the MOS Transistor, Part 3," *Electronic Engineering Journal,* April 17, https://www.eejournal.com/article/a-brief-history-of-the-mos-transistor-part-4-ibm-research-persistence-and-the-technology-no-one-wanted; Intel, 2024, "About Intel Israel," https://www.intel.com/content/www/us/en/corporate-responsibility/intel-in-israel.html; Applied Materials, Applied Materials Finland, https://www.appliedmaterials.com/eu/en/about/europe-overview/finland-overview.html.

[16] See DoD, 2020, "Department of Defense Announces $197.2 Million for Microelectronics," October 15, https://www.defense.gov/News/Releases/Release/Article/2384039/department-of-defense-announces-1972-million-for-microelectronics; Intel, 2023, "RAMP-C Program on Intel 18A Adds 2 Strategic Defense Industrial Base Customers," July 18, https://www.intel.com/content/www/us/en/newsroom/news/ramp-c-program-intel-18a-adds-strategic-defense-industrial-base-customers.html.

architecture fabrication process or facility.[17] In the 2023 RAMP-C program, Intel's foundry program enabled two leading DoD systems suppliers, Boeing and Northrop Grumman, to use Intel's upcoming 18A process technology and industry-standard design tools and intellectual property (IP) to develop and fabricate customized test chips for DoD in preparation for product insertions. Previously, RAMP-C included arrangements with Nvidia, Qualcomm, IBM, and Microsoft. DoD's related State-of-the-Art Heterogeneous Integrated Packaging (SHIP) program focuses on packaging for prototype devices.[18] These three programs—RAMP, RAMP-C, and SHIP—created innovative DoD contractual arrangements with companies to meet DoD microelectronics needs.

Occasionally, commercial customers for leading-edge semiconductors have entered into partnership agreements with semiconductor manufacturers, investing in those companies to ensure they have guaranteed access to needed chip supply to meet their business needs. Recent examples include the following: Tower Semiconductor making a capacity investment in an Intel fab line in New Mexico; Vitesco paying for new silicon carbide production capacity at an OnSemi fab; Qualcomm financing a capacity expansion by GlobalFoundries; and Sandisk investing in Toshiba capacity expansion.[19] In these cases, customers were investing in semiconductor manufacturer capacity to ensure that their needs for manufactured semiconductor products would be met. There is no fundamental reason why DoD cannot fund a similar arrangement that serves the same essential purpose—ensuring that needed capacity to build leading-edge semiconductors for DoD use is available in the United States.

---

[17] DoD, 2024, "Rapid Assured Microelectronics Prototypes Using Advanced Commercial Capabilities (RAMP) Project," https://www.cto.mil/ramp-project.

[18] DoD, 2023, "Department of Defense Celebrates Advancements in Microelectronics Capabilities," April 6, https://www.defense.gov/News/Releases/Release/Article/3355049/department-of-defense-celebrates-advancements-in-microelectronics-packaging-cap.

[19] Intel, 2023, "Intel Foundry Services and Tower Semiconductor Announce New Foundry Agreement," Press Release, September 5, https://www.intc.com/news-events/press-releases/detail/1643/intel-foundry-services-and-tower-semiconductor-announce-new. For other examples, see *Business Wire*, 2023, "Bitesco Technologies and Onsemi Sign Long Term Supply Agreement," May 31, https://www.businesswire.com/news/home/20230531005876/en/Vitesco-Technologies-and-onsemi-Sign-SiC-Long-Term-Supply-Agreement-and-Agree-to-Invest-in-SiC-Technology-Capacity-Expansion; *The Register*, 2022, "GlobalFoundries Double Down on US Chip Production," *Qualcomm*, August 8, 2022, https://www.theregister.com/2022/08/08/globalfoundries_qualcomm_us; "AnandTech, Toshiba Memory and Western Digital Finalize Fab K1," May 10, 2019, https://www.anandtech.com/show/14359/toshiba-memory-western-digital-finalize-fab-k1-investment-agreement; The Street, 2006, "Toshiba to Build Japanese Chip Plant," SanDisk, August 4, https://www.thestreet.com/technology/sandisk-toshiba-to-build-japanese-chip-plant-10301922; Toshiba News Release, 2010, "Toshiba and SanDisk Sign Joint Venture Agreement," July 14, https://www.global.toshiba/ww/news/corporate/2010/07/pr1401.html.

There are a variety of mechanisms that the government has historically used to subsidize the availability of capacity in the United States to meet potential defense needs in an emergency, including loan guarantees, air carrier capacity assurance, the Defense Production Act, and the government-owned, contractor-operated (GOCO) approach. These mechanisms are summarized below, and may have potential applicability to meeting DoD semiconductor needs.

- *Loan guarantees.* Loan guarantees from the government provide a form of subsidy to capital investments. The committee understands that this authority is under consideration in DoD through its new Office of Strategic Capital. If this new office develops its lending capability at sufficient scale, it may allow DoD to access semiconductor manufacturing capacity as needed.
- *Air carrier capacity assurance.* Under the Civil Reserve Air Fleet program, domestic commercial airlines contractually commit aircraft for potential DoD emergency use in return for peacetime DoD airlift business directed to the participating airlines.[20] Here, DoD leverages its ongoing spending on air transport through commercial air carriers to also assure a reserve air transport fleet. A conceptually similar approach to DoD's semiconductor needs could be envisioned, although the current relatively low volume of DoD semiconductor purchases likely will not, by itself, yield enough leverage to motivate private chipmakers to guarantee later capacity access.
- *Defense Production Act.* The Defense Production Act of 1950 (DPA) is the legal foundation for application of U.S. commercial industrial might to DoD needs from the 1950s through the present day. It is a frequently wielded tool that DoD has used to align U.S. technology and industrial investments with national security needs in recent decades.[21] The DPA created a framework

---

[20] U.S. Air Force, 2014, "Civil Reserve Air Fleet (CRAF)," July 2014, https://www.af.mil/About-Us/Fact-Sheets/Display/Article/104583/civil-reserve-air-fleet.

[21] The U.S. Army first used procurement contracts to fund development of new manufacturing technology in the period after an early "national security shock," the War of 1812. Responding to problems in securing supplies of needed firearms as European imports, the Army at first gave contracts to U.S. start-up firms to develop technology for the high-volume production of firearms with interchangeable parts, investing in facilities to mass produce them. These contracts typically advanced funds to contractors (initially, Eli Whitney's factory in North Haven, Connecticut) to cover the development costs, as well as dual-use factory capital investments, that could then be utilized to sell these products to the government at negotiated prices. The Army's arsenals at Harper's Ferry, West Virginia, and Springfield, Massachusetts, took over the development R&D after Whitney's efforts proved incomplete. Colt, Remington, and Smith and Wesson were all subsequent famous U.S. firearms start-ups that commercialized these products based on the interchangeable parts approach and the machine tools behind it. The new process initiated the creation of the "American system of manufactures," homegrown industrial technology that powered U.S. manufacturing industries to global prominence in the 19th century and early 20th centuries. D. Hounshell, 1984, *From the American System to Mass Production, 1800–1932: The Development of Manufacturing Technology in the United States*, Johns Hopkins University Press; M. Roe Smith (ed.), 1985, *Military Enterprise and Technological Change*, MIT Press; M. Roe Smith, 1977, *Harpers Ferry Armory and the New Technology*, Cornell University Press.

that permits co-funding investments in industrial technology and production capacity investment for dual-use goods within the structure of DoD procurement contracts. DoD might consider running a competition for procurement contracts that would provide payment of some percentage of certain costs—for example, in a first phase, the costs of developing a next-generation, leading-edge fabrication technology node; in a second phase, the costs of building a commercial foundry capability in the United States. In exchange for DoD covering a percentage of the cost, the U.S. government would be contractually guaranteed priority access for up to that percentage of the capacity of the leading-edge industrial facility implementing this technology, on commercial terms. In normal times, the commercial firms would be authorized to sell unutilized defense foundry capacity to ordinary commercial users on normal terms. Outside of national security emergencies, the government would also pay for capacity on the same terms as commercial users. However, in times of emergency or urgent national security need, such foundries would be obligated contractually to accept Defense Production Allocation System (DPAS) DX- and DO-rated[22] priority orders, on commercial terms, for up to the set percentage of the foundry's production capacity (in new wafer starts), throughout the economic life of that foundry technology node. Capacity subsidy payments could be structured as progress payments on the relevant procurement contracts. Progress payments are an advanced financing tool that DoD has often used to fund big-ticket, risky, new technology and systems investment contracts intended to create innovative new defense systems.

- *GOCO approach.* DoD currently operates many GOCO facilities. The U.S. Air Force, for example, owns several aircraft manufacturing plants in which U.S. government facility investments are comingled with private contractor investments in production equipment and tooling. Examples include Air Force plants operated by Lockheed Martin in Fort Worth, Texas, and Marietta, Georgia (AF Plant 4 and AF Plant 6), where the F-35, F-16, F-22, C-6, C-130, and P-3 aircraft are produced, overhauled, or maintained. In addition to selling these aircraft to the U.S. government, the private contractors operating these facilities are permitted to make commercial sales to other customers.[23] The U.S. Army, in order to meet its need for ammunition plants, has also built plants, then contracted with private companies to operate these Army-owned plants. The company makes the product for commercial

---

[22] DO-rated orders require the signature of the Under Secretary of Defense in charge of acquisition and technology; DX-rated orders require the signature of the Secretary of Defense. The order of priority is DX orders, then DO orders, then commercial. See Defense Contract Management Agency (DCMA), Defense Priorities and Allocation System (DPAS), Rating System, https://www.dcma.mil/DPAS.

[23] The government reportedly owns about 80 percent of the industrial facility floor space in these two plants, with the remainder, along with the industrial equipment and tooling, owned by the private contractor. See GlobalSecurity.org, Military, Air Force Plants 4 and 6, https://www.globalsecurity.org/military/facility/afp-4.htm, and https://www.globalsecurity.org/military/facility/afp-6.htm.

sales and makes modernization investments, but also supplies the military. The military, as plant owner, can preempt commercial sales to meet any emergency "surge" requirements.[24] The U.S. government has also invested in GOCO partnerships with private contractors to meet other national needs. Private contractors operate 16 out of 17 Department of Energy National Laboratories and are encouraged to make the capacity and capabilities of these laboratories available for sale to non-government customers. The National Institutes of Health (NIH) owns a private contractor–operated vaccine production facility that sells vaccines to private entities conducting clinical trials.[25] Because it is difficult to visualize a scenario in which DoD needs would ever amount to more than a small fraction of the output of a commercial volume foundry, majority government ownership seems excessive. However, an arrangement with the government owning a minority share, in exchange for access to capacity if needed, appears a reasonable arrangement.

Given that guaranteed supply of advanced semiconductors for use in U.S. military systems is arguably as critical to a 21st-century military as ammunition rounds were in the 20th century, all of these partnership mechanisms should be considered to assure DoD can obtain the chips it needs.

Leveraged Access for Department of Defense to Fabrication Plants

Semiconductor fabs only remain state-of-the-art for 3–5 years, until superseded by the next generation of advanced technologies and newer fabs. A lesson from the past two decades of semiconductor manufacturing is that advanced fabs will only be constructed in the United States if federal government financing matches that of competitor nations—the essential approach taken by the CHIPS Act. As suggested above, DoD could benefit from a collaborative agreement to meet its advanced chip needs as part of such government financing. That is, the government might finance a share of the new process technology development and fab construction in exchange for DoD obtaining guaranteed access to a share of the resulting new capacity. The U.S. government could combine a strategy for assuring U.S.-based manufacturing of advanced chips with a strategy for assuring DoD's access to these facilities. (See Recommendation 5.13.)

---

[24] See, for example, L. McLaughlin, n.d., "Government-Owned, Contractor Operated 101," BAE Systems, https://www.baesystems.com/en-us/feature/government-owned-contractor-operated-101.

[25] PrEP4All, 2022, "Deploying the Government Owned, Contractor Operated Model: How the U.S. Government Can Stand Up a Billion Dose mRNA Vaccine Facility Within Six Months," *Public Citizen*, March 8, https://www.citizen.org/article/deploying-the-government-owned-contractor-operated-model.

OVERCOMING BARRIERS TO LOWER-COST DESIGN

The soaring costs of chip design for ASICs are a barrier to DoD using the highly integrated, customized chips to create technological superiority in military systems. This is the third category of chip discussed in the DSB's 2006 report, namely, high-performance chips designed specifically for insertion into military systems to achieve technological advantage. This is not a new challenge: even in 2006, design costs for new custom chips were soaring, although not to the levels now seen. The 2006 DSB report's very first recommendation was for an initiative to develop new design tools "to enable the design of affordable low volume, high performance custom ASICs."[26] The logic driving that recommendation is even more relevant today.

In 2006, the DSB recommended that tool development

> should be integrated into the industry mainstream toolset for CMOS design. The development can be cost-shared with the commercial industry. The commercial Electronic Design Automation industry is currently developing such tools, but rate of investment is not sufficient to meet the challenge for DoD. While commercial industry will benefit from these tools, DoD will benefit disproportionately because of the importance of their contribution to the development of low volume, high performance custom ASICs.

These observations remain true today. What has changed is that advances in software and artificial intelligence technology make very rapid progress more plausible today than it was 18 years ago.

The CHIPS Act focused on production facilities rather than on design costs. Significantly lowering design costs remains a critical barrier to DoD exploiting technology opportunities that would result from access to the most advanced chip manufacturing technology. Furthermore, DoD is not alone on this issue. Design cost at the most advanced technology nodes is a major challenge faced by start-ups and small and midsized firms, and large firms trying to tailor applications to relatively moderate-sized markets using the most advanced chips. Overall, many parts of industry have a significant interest in pursuing lower-cost design technologies. Thus, a collaborative R&D initiative led by DoD working with industry to develop lower-cost design approaches would also have broadly beneficial commercial impacts.

It is evident that human resource constraints increase costs and limit the insertion of advanced system semiconductors into new DoD systems. Use of AI and other tools may reduce the human input required for new advanced chip design and development perhaps by an order of magnitude. That is, the 300 to 400 software engineers required on a complex chip development effort potentially could be reduced to 30 or 40, and the 100 required on a simpler project could be reduced

---

[26] DSB, 2006, "Joint Task Force on Critical Technologies," p. 69.

to 10. Overall, more efficient and lower-cost design represents a major technology challenge and DoD, in cooperation with the National Institute of Standards and Technology (NIST) and the National Semiconductor Technology Center (NSTC), would be well served by tackling this challenge. Progress in design automation will also assist in addressing the workforce training and supply issues discussed in Chapter 7. Furthermore, given the significant potential expected from advances in design automation, if the United States does not undertake this effort, it may well cede its current leadership in chip design to other nations. (See Recommendations 5.5, 5.6, and 5.7.)

OVERCOMING RESEARCH AND DEVELOPMENT BARRIERS TO NEXT-GENERATION SEMICONDUCTORS

In the R&D area, significant changes may be ahead for the semiconductor sector.[27] The effort to decrease CMOS-based[28] chip feature sizes has lasted for decades, but with node sizes reaching 3 nm, and potentially 2 or 1.5 nm, this approach is becoming increasingly difficult.[29] The need for post-CMOS technologies may arrive within the next decade.[30]

New approaches involving different kinds of architectures, materials, packaging, software, and algorithms will be required. In general, the semiconductor industry needs to pursue three lines of effort to realize continued performance improvements for microelectronics: (1) more efficient architectures and packaging, (2) new models of computation, and (3) new materials and devices.[31] The IEEE's International Roadmap for Devices and Systems (IRDS) has identified a series of beyond-CMOS approaches, including the following: computational state variability, nonthermal equilibrium systems, novel energy transfer interactions, nanoscale thermal management, sublithographic management, and alternative architectures.[32] Architecture advances that appear promising include in-memory

---

[27] Office of Science and Technology Policy (OSTP), 2024, "National Strategy for Microelectronics Research," March, pp. 7–17.

[28] Complementary metal-oxide-semiconductor (CMOS) is a type of fabrication process for logic functions used for integrated circuit chips and first developed in 1963 and introduced in 1986 by Intel. See history in B. Lojek, 2007, *History of Semiconductor Engineering*, Berlin: Springer.

[29] M. Lapedus and E. Sperling, 2020, "Making Chips at 3nm and Beyond," *Semiconductor Engineering*, April 16, https://semiengineering.com/making-chips-at-3nm-and-beyond.

[30] C. Leiserson, N. Thompson, J. Emer, B. Kuszmaul, B. Lampson, D. Sanchez, and T. Schardl, 2020, "There's Plenty of Room at the Top: What Will Drive Computer Performance After Moore's Law?," *Science* 368:6495, https://www.science.org/doi/10.1126/science.aam9744.

[31] J. Shalf, 2020, "The Future of Computing Beyond Moore's Law," *Royal Society*, https://royalsociety publishing.org/doi/10.1098/rsta.2019.0061.

[32] Institute of Electrical and Electronics Engineers (IEEE), 2018, "Beyond CMOS: The Future of Semiconductors," IEEE IRDS, https://irds.ieee.org/home/what-is-beyond-cmos.

processing (to improve the power usage and performance of moving data between processor and memory), nano/micro electro-mechanical systems (N/MEMS), magneto-electric spin-orbit (MESO) logic devices, cryogenic electronics (low-temperature electronics), structures based on magnetic tunnel junctions, and edge computing (which processes data close to where it is generated, enabling greater speed and volume).[33] Use of 3D heterogeneous integration (3DHI) packaging (technologies that stack numerous integrated circuits and interconnect them vertically for improved performance at lower power and smaller size) is currently receiving attention, as described in more detail below. The 2022 report by DoD, "Future Directions Workshop: Materials, Processes, and R&D Challenges in Microelectronics," emphasizes the need to explore scientific research in such areas as new materials for design approaches that involve alternatives to electron transport (including photonics, spintronics, topological materials, and exotic quasi-particles). In addition, new engineering processes could provide new materials approaches (including 3D integration, atomic scale fabrication and metrologies, and digital twins for semiconductor processes and microarchitectures). The report suggests new opportunities for post-CMOS advances.[34] (See Recommendations 5.2 and 5.3.)

During the 1950s and 1960s, faced with an apparent peer competitor, DoD made large investments in its industrial base and took risks in creating new weapons systems. DoD (and NASA)[35] encouraged its contractors to incorporate newly invented electronic devices into its systems even when reliability and performance were somewhat uncertain. Arguably, the security climate today is more like the 1960s than the 1990s, and DoD would be justified in absorbing a little more risk in exchange for increasing the odds of improvements in capability and quality. Advances in chip density, speed, and energy use are vitally important to DoD (as well as, of course, to many of the industries dependent on semiconductors). Industry leaders are focused more on the advances needed in the next decade, and less on subsequent technologies needed to maintain progress over the long term. DoD could play an important role in the coordination and support of these next-generation R&D efforts with industry, especially if funding for NSTC is not authorized past the initial 5 years.

An ongoing commitment to post-CMOS R&D appears in order; the NSTC may start to satisfy that need for the 5 years it has assured funding, and DoD should

---

[33] See also approaches detailed in DoD Basic Research Office, 2022, "Future Directions Workshop: Materials, Processes, and R&D Challenges in Microelectronics," Department of Defense, June 23–24, https://basicresearch.defense.gov/Portals/61/Documents/future-directions/FDW%20Microelectronics%20Report_with%20DOI.pdf.

[34] S. Guha, H.S.P. Wong, J.A. Incorvia, and S. Chowdhury, 2022, "Future Directions Workshop: Materials, Processes, and R&D Challenges in Microelectronics," Department of Defense, June 23–24.

[35] C. Fishman, 2020, *One Giant Leap*, New York: Simon and Schuster.

participate fully to ensure its concerns are met. DoD should also consider a separate effort for the longer-term support required and for its needs, in cooperation with the NSTC, building on the R&D elements within DARPA's Electronic Resurgence Initiative (ERI) 2.0.[36]

Leading-edge silicon semiconductor manufacturing technology is important because it is the hardware platform onto which new information technology (IT)—including not only new kinds of electronic components, but also new software and algorithms, photonics, AI, and quantum IT—is going to be incorporated into systems. However, metrics for progress in silicon MOS fabrication technology, such as the pace of improvement in speed, energy consumption, size, and density, have slowed relative to historical norms. The semiconductor device fabrication industry seems to have matured relative to its state just a decade or two in the past. Investments in software, AI, quantum information systems, and photonics may have both greater uncertainty and higher potential payoffs. It is important that DoD be positioned to take advantage of the latter.

In March 2024, the Subcommittee on Microelectronics Leadership of the National Science and Technology Council in the Office of the President released a strategy paper on microelectronics research.[37] The strategy focused on four areas: (1) accelerating research on future generations of advances on microelectronics; (2) building the research infrastructure, including tools and equipment, that can lead to prototyping and subsequent fabrication, assembly, packaging, and testing of new advances; (3) supporting the workforce that will be required both for research and implementation; and (4) creating the networks and activities that will support a robust innovation ecosystem in microelectronics. At the heart of the strategy is an aggressive pursuit of new R&D; the other three recommendations constitute supporting pillars for this effort. The R&D approach is multifaceted, calling for development of new processes and metrology for advanced packaging and integration; semiconductor materials that enhance capabilities and functionality; circuit design, simulation, and emulation tools; processing architectures and related hardware for future systems; manufacturing tools and processes for fabrication—all while prioritizing security as an element in design strategies. It also calls for improved access for start-ups, small firms, and university researchers to advanced facilities where they can prototype devices and processes. The issues identified in this report span both the needed short-term advances in packaging and integration as well as the longer-term advances listed in the National Science and Technology Council report, which can be generally described as post-CMOS

---

[36] DARPA, 2023, "Electronics Resurgence Initiative 2.0," March 6, https://www.darpa.mil/work-with-us/electronics-resurgence-initiative.

[37] OSTP, 2024, "National Strategy for Microelectronics Research," White House, March, https://www.whitehouse.gov/wp-content/uploads/2024/03/National-Strategy-on-Microelectronics-Research-March-2024.pdf.

technologies that are likely to be needed over the next decade. The recommendations that follow echo many of the specifics identified in the National Science and Technology Council strategy.

PACKAGING, INTEGRATION, AND OTHER NEAR-TERM TECHNOLOGIES

Progress is well under way on a set of semiconductor technology advances that could be realized in approximately 5 years. At the top of the list are efforts on packaging and integration, backed by both DARPA and NIST.

In late 2023, DARPA announced plans for a new center for Next-Generation Microelectronics Manufacturing (NGMM) to deliver new approaches for microelectronics fabrication. The center was initially funded at $430 million and is intended to undertake both research and prototyping, with a focus on the challenges of tomorrow and operating over longer timelines than efforts funded under the CHIPS Act. The premise is that by integrating and packaging chip components differently via 3DHI, manufacturers will be able to significantly improve performance. DARPA states:

> The end goal of the program is to establish a self-sustaining 3DHI manufacturing center at an existing facility that is owned and operated by a non-federal entity and accessible to users in academia, government, and industry. Success will be measured by the ability to produce high-performance 3DHI microsystems at reasonable cost, with cycle times supporting fast-paced innovative research.[38]

The NGMM initiative recognizes that the semiconductor industry is shifting from mass-producing increasingly complex monolithic chips to creating chiplets that can be packaged together in three dimensions for customized applications.[39] According to Pat Gelsinger, CEO of Intel, "Essentially, old silicon stuff is becoming cool new packaging stuff. We're now entering into the advanced-packaging era, where 2-and-a-half and 3D packages become the new norm."[40] Recognizing the potential of this transition, Mark Rosker, director of DARPA's Microsystems Technology Office, indicates the NGMM program is "looking at ways to incorporate photonics and non-silicon electronics into these systems."[41]

---

[38] DARPA, 2023, "Next-Generation Microelectronics Manufacturing Aims to Sustain R&D Ecosystem: DARPA Names Eleven Teams to Kick Off Groundbreaking U.S. Chips-of-the-Future Effort," July 20.

[39] Y. Jie, 2023, "To Drive AI, Chip Makers Stack 'Chiplets' Like Lego Blocks: Nvidia, Intel, AMD and Others Invest in Technology That Promises More-Powerful, Easier-to-Build Semiconductors," *Wall Street Journal*, July 10.

[40] A. Boyle, 2023, "DARPA and Other Federal Agencies Working on Strategies to Revive America's Chip Industry," *GeekWire*, August 22, https://www.geekwire.com/2023/darpa-summit-revive-chip-industry.

[41] Ibid.

NIST, too, is embarking on a major packaging and integration initiative with its National Advanced Packaging Manufacturing Program (NAPMP), funded with $3 billion from the CHIPS Act.[42] Elements in the NAPMP program include materials and substrates; advances in tools, equipment, and processes; power delivery and thermal management; photonics and connectors; the chiplet ecosystem; and design using automated tools and built-in test and repair. Coordination of the DARPA and NIST programs will be important because elements of the two programs overlap.

DoD's recently formed Microelectronics Commons (ME Commons), comprising a set of public–private partnerships (PPPs) to enable semiconductor technology transitions from laboratory to fabrication, to improve U.S. leadership in semiconductor technical development, and to accelerate the transition of new semiconductor technologies to domestic producers.[43] This 5-year CHIPS and Science Act program is aimed at six DoD semiconductor technology priorities: 5G and 6G broadband technology, AI and hardware, commercial leap-ahead technologies, electromagnetic warfare, Internet of Things and secure-edge computing, and quantum technology. It also aims to better connect university researchers in semiconductor fields with semiconductor production facilities to enable further research advances. Because of its somewhat shorter-term technology adoption focus, ME Commons is not positioned to address longer-term technologies.

NIST is forming the NSTC as a PPP consortium for semiconductor R&D and will engage academia and industry around barriers to technology advancement in the semiconductor industry, including workforce needs. The consortium will be a separate nonprofit corporation built around the consortia. The key strategic goals for the NSTC are to (1) extend U.S. leadership in foundational technologies for future applications and industries and strengthen the U.S. semiconductor manufacturing ecosystem, (2) significantly reduce the time and cost to prototype innovative ideas, and (3) build and sustain a semiconductor workforce development ecosystem.[44] The NSTC has the same 5-year timeline as other CHIPS Act programs.

---

[42] NIST, 2023, "The Vision for the National Advanced Packaging Manufacturing Program," CHIPS R&D Office, November 20, https://www.nist.gov/system/files/documents/2023/11/19/NAPMP-Vision-Paper-20231120.pdf. This packaging initiative was funded in the CHIPS and Science Act in 2022. See U.S. Senate Commerce Committee, The CHIPS Act of 2022, Section by Section Summary, https://www.commerce.senate.gov/services/files/592E23A5-B56F-48AE-B4C1-493822686BCB.

[43] DoD, 2024, "Microelectronics Commons," https://microelectronicscommons.org.

[44] NIST, 2024, "National Semiconductor Technology Center," July 12, https://www.nist.gov/chips/research-development-programs/national-semiconductor-technology-center. See also NIST, 2023, "NSTC Update to the Community," November 16, https://www.nist.gov/chips/national-semiconductor-technology-center-update-community, and NIST, 2023, "A Vision and Strategy for the NSTC," April 25, https://www.nist.gov/system/files/documents/2023/04/27/A%20Vision%20and%20Strategy%20for%20the%20NSTC.pdf.

Although NSTC's existing appropriation is multiyear and can be stretched beyond 5 years, it is not clear whether that appropriation will be extended after that period, despite the expected needs. Because of their timelines, both the NSTC and ME Commons may naturally tend to focus on technologies that can be implemented in their 5-year periods.

The Need for Coordination Between the National Institute of Standards and Technology and the Department of Defense Near-Term Research and Development Programs

The establishment of DoD's NGMM Center and ME Commons as well as NIST's NAPMP and NSTC should enable substantial progress on the interim R&D agenda for semiconductor advances, including packaging and integration as well as other technologies that should be available for implementation in the coming half decade. It is vital that all these programs be effectively coordinated to assure exchange of ideas and avoid duplication and research dead ends. (See Recommendation 5.1.)

These programs should also assist in lowering another barrier to progress in semiconductor research—the need to provide university semiconductor researchers as well as start-up companies with access to advanced fabrication facilities. (See Recommendation 5.2.)

POST-COMPLEMENTARY METAL-OXIDE-SEMICONDUCTOR ADVANCES

In addition to the potential initial advances noted above, DoD needs to consider longer-range semiconductor technology advances that could evolve in the coming decade. Those are centered around post-CMOS technologies to achieve needed gains in speed, energy use, and density. As the semiconductor sector's International Roadmap for Devices and Systems (IRDS) has found, "the future of the semiconductor industry is headed Beyond CMOS."[45] Unless a systematic effort commences relatively soon, those technologies will not be accessible in approximately 10 or so years as the current scaling efforts play out.

Major gains are possible through a wide range of possible CMOS approaches, many of which are listed at the outset of this section on R&D barriers, as identified in the IRDS. This range gives researchers flexible options to pursue to come up with promising approaches. The Nanoelectronics Research Initiative (NRI) of the Semiconductor Research Corporation (SRC) is attempting to benchmark these potential

---

[45] IEEE, 2023, "Beyond CMOS: The Future of Semiconductors."

approaches against each other.[46] There are also significant efficiency gains possible through software performance engineering, which entails restructuring software so computer applications can run more quickly by removing unneeded software elements and tailoring software to specific features of the hardware architecture.[47] Gains are also possible through new algorithms with lower-levels of computation, as well as through hardware streamlining that implements functions with fewer transistors and less chip area. While there are currently two basic forms of chips, CPUs (central processing units, the logic chip) and GPUs (graphics processing units, applicable to computer graphics and AI uses), more domain-specific, specialized chips could well evolve, bringing performance gains to specific application areas.

Initiating a Broad Post-Complementary Metal-Oxide-Semiconductor Research and Development Effort

While NIST's NSTC, and to some extent DoD's ME Commons, may include research on some of these areas, a systematic effort is needed to accelerate research on post-CMOS technologies and approaches that meet commercial and DoD needs, working with NSTC and NIST. (See Recommendations 5.2 and 5.3.)

OVERCOMING REGULATORY BARRIERS

DoD faces a series of barriers from internal security requirements that affect its ability to rapidly adopt advanced chips.

To regulate the security of its platforms and systems, DoD in 2004 developed a "trusted foundry" program initially with IBM and subsequently with other companies for assured, secure manufacturing at U.S. plants of the chips it requires.[48] The goal of this program was to ensure that DoD understood and controlled the provenance of all custom microelectronics it was procuring. The program reaches integrated circuit design, aggregation, mask making, foundry, and packaging steps to try to assure a "chain of custody" for both classified and unclassified integrated circuits. For example, DoD announced a $3.1 billion 10-year agreement with GlobalFoundries to meet semiconductor needs through the trusted access program.

---

[46] Semiconductor Research Corporation (SRC), 2024, "Nanoelectronics Research Initiative (NRI)," https://www.src.org/program/nri; A. Chen, 2016, "An Overview of Nanoelectronics Research Initiative (NRI)," Paper presented at the 2016 13th IEEE International Conference on Solid-State and Integrated Circuit Technology, *ICSICT 2016 – Proceedings.*

[47] C.E. Leiserson et al., 2020, "There's Plenty of Room at the Top: What Will Drive Computer Performance After Moore's Law?" *Science*, June 5, https://www.microsoft.com/en-us/research/uploads/prod/2020/11/Leiserson-et-al-theres-plenty-of-room-at-the-top.pdf.

[48] Defense Microelectronics Activity (DMEA), 2024, "DMEA Trusted IC Program," https://www.dmea.osd.mil/TrustedIC.aspx.

GlobalFoundries has dropped out of the race to produce the most advanced chips, however, producing chips at or above the 12 nm scale.[49] Because semiconductors are pervasive in DoD's equipment, as noted in Chapter 2, the great majority of the semiconductors embedded in DoD systems are commercial-off-the-shelf components. Only a small minority are customized chips to meet specific DoD functionality, performance, or security requirements. While DoD's trusted suppliers potentially could meet the latter need, the system does not embrace the most advanced chips, and only 2 percent of DoD's chips are from that trusted supply chain system.[50]

This has resulted in challenging economics for trusted foundry participants: companies that become "trusted" undergo significant effort to align their business with DoD requirements, and frequently the volume of DoD trusted foundry purchases cannot generate the return on investment these firms need to remain suppliers to DoD's Trusted Foundry program. The Trusted Foundry program offers a limited suite of microelectronics for DoD systems when evaluated by technology and node.[51] Although the Trusted Foundry program has roughly 80 facilities accredited, only 16 of these facilities are fab lines.[52] Of these fab lines, many are sole-source suppliers. In addition, access to GlobalFoundries's 12 nm fab line noted above (the most advanced in the trusted foundry ecosystem), which was recently accredited,[53] to date has been quite limited for wafer production runs and untested wafers.[54]

Given these ongoing challenges, in May 2020, the director of defense research and engineering for modernization within the Office of the Under Secretary of Defense for Research and Engineering (OUSD R&E) stated that the current

---

[49] K. Tarsov, 2023, "How GlobalFoundries Aims to Remain World's Third Biggest Semiconductor Foundry," *CNBC*, October 1, https://www.cnbc.com/2023/10/01/how-globalfoundries-aims-to-remain-worlds-third-biggest-chip-foundry.html. Of course, there is a strategic aspect to legacy chips as well as leading-edge chips, because many industries, such as automotive, as well as DoD systems, rely on them. See S. Shivakumar, C. Wessner, and T. Howell, 2023, "The Strategic Importance of Legacy Chips," Center for Strategic and International Studies, March 3, https://www.csis.org/analysis/strategic-importance-legacy-chips.

[50] J.M. Donnelly, 2021, "Pentagon Races to Shore Up Supply Chain Security," *Government Technology*, April 9, https://www.govtech.com/security/pentagon-races-to-shore-up-supply-chain-security.html.

[51] DMEA, 2024, "Defense Microelectronics Activity (DMEA): TAPO—Foundry Services," https://www.dmea.osd.mil/TAPOFoundry.aspx.

[52] DoD, 2023, "Accredited Suppliers," Trusted Foundry Program, Defense Microelectronics Activity, September 7, https://www.dmea.osd.mil/otherdocs/AccreditedSuppliers.pdf.

[53] GlobalFoundries, 2023, "GlobalFoundries and U.S. Department of Defense Showcase Secure Chip Manufacturing at 2023 Trusted Foundry Training," October 12, https://gf.com/gf-press-release/globalfoundries-and-u-s-department-of-defense-showcase-secure-chip-manufacturing-at-2023-trusted-foundry-training.

[54] DMEA, 2024, "Defense Microelectronics Activity (DMEA): TAPO—Foundry Services."

method for acquiring custom microelectronics from a trusted supplier (trusted foundry) system had failed.[55] The director further stated that to access state-of-the-art (SOTA) microelectronics, DoD needed to move to an evidence-based assurance method that can leverage commercial industry while maintaining hardware security (like Microelectronics Quantifiable Assurance [MQA], discussed below).[56]

The trusted foundry system is part of the Defense Microelectronics Activity (DMEA), a regulatory and testing entity with some 200 staff based in McLellan, California. It manages the trusted access program; the advanced technology support program for microelectronics acquisition and oversight; an engineering program for developing microelectronics solutions, including for legacy systems; and radiation testing for delivering radiation-hardened and tested microelectronics.[57] DMEA also manages its own foundry, known as the Advanced Reconfigurable Manufacturing for Semiconductors (ARMS) facility.[58]

The regulatory underpinnings for this system run through the International Traffic in Arms Regulations (ITAR) and Export Administration Regulations (EAR), which regulate the export of defense-related technologies, information, and services, including information conveyed to (and therefore work performed by) nonpermanent residents.[59] The Directorate of Defense Trade Controls in the Department of State leads ITAR, and the Bureau of Industry and Security in DOC leads EAR. ITAR regulates semiconductors and related equipment and materials used in military and space applications, and EAR regulates devices that are dual-use and can be used in either civilian or defense applications. Although much of the talent base for semiconductor R&D is foreign-born, ITAR limits their involvement in government-supported R&D or work, affecting the ability to bring the best talent

---

[55] DoD Inspector General, 2022, "Evaluation of the Department of Defense's Transition from a Trusted Foundry Model to a Quantifiable Assurance Method for Procuring Custom Microelectronics (DODIG-2022-084)," May 4, https://www.dodig.mil/reports.html/Article/3019461/evaluation-of-the-department-of-defenses-transition-from-a-trusted-foundry-mode.

[56] Ibid. See also Air Force, 2023, "AA2023 Mandated Independent Review of Microelectronics Quantifiable Assurance," August 3, https://www.af.mil/Portals/1/documents/2023SAF/MQA_Report.pdf.

[57] DMEA, 2024, "General Information," https://www.dau.edu/acquipedia-article/defense-microelectronics-activity-dmea.

[58] A March 2020 DoD Inspector General report found that DMEA spent $32.4 million between January 2014 and July 2019 to maintain the ARMS foundry while only using it to provide five wafer lots for five DoD customer requests, calling into question its utility and return on investment. But $30 million may be a modest price tag if it enables an important legacy platform to continue to operate. DoD Inspector General, 2020, "Audit of DoD Hotline Allegations Concerning the Defense Microelectronics Activity (DODIG-2020-072)," March 26, https://www.dodig.mil/reports.html/Article/2126034/audit-of-dod-hotline-allegations-concerning-the-defense-microelectronics-activi.

[59] Code of Federal Regulations (CFR), Title 22, Chapter 1, Subchapter M, Part 120, https://www.ecfr.gov/current/title-22/chapter-I/subchapter-M.

to work on critical advances. A 2020 study found that 40 percent of high-skilled semiconductor workers in the United States were born abroad.[60] These regulations involve two additional agencies, creating additional incremental delays in serving DoD's needs for chips. (See Recommendation 5.10.)

The inefficiency of the current system for securing custom advanced chips exacts a significant toll on national security readiness. The recent experience of the Air Force in developing a new jam-resistant Global Positioning System (GPS) capability, a vital defense strategy as the recent war in Ukraine illustrates, was a billion-dollar project, and the needed ASIC chip was designed in a short amount of time. However, it took years of maneuvering through administrative barriers justifying utilization of U.S. facilities at one of the two U.S. suppliers, GlobalFoundries or Intel, which were not even fabricating the most advanced chips available globally. If DoD is effectively blocked by administrative barriers from timely access to advanced chips, then the national security justification behind the CHIPS Act is undermined. It will be important for future procurements of advanced chip designs to benefit from this experience through streamlined procedures that enable much faster adoption.

To utilize the most advanced chips for its custom needs, DoD increasingly must rely on and embrace commercial electronics supply chains. This is at odds with DoD's traditional approach to security, which historically emphasized cradle-to-grave production in the continental United States, export controls on technology, security through obscurity,[61] and numerous military standards (which often differ substantially from commercial practices) for interfaces, design, manufacturing, standards and testing of electronic devices, and the associated manufacturing processes. Of course, numerous exceptions have been authorized to permit DoD to meet its needs, but significant restrictions on design and manufacturing preferences for custom chips remain. DoD's continuing restrictions place it at odds with practices in the commercial microelectronics industry, which is inherently global, collaborative, strongly committed to a set of commercial standards, and focused on low-mix, high-volume manufacturing. DoD faces a trade-off: it cannot secure access to today's leading-edge microelectronics without altering its system for how it manages a secure supply chain. The more closely DoD can align with the

---

[60] W. Hunt and R. Zwetsloot, 2020, "The Chipmakers. U.S. Strengths and Priorities for the High End Semiconductor Workforce," Center for Security and Emerging Technology (CSET), Georgetown University, September, https://cset.georgetown.edu/publication/the-chipmakers-u-s-strengths-and-priorities-for-the-high-end-semiconductor-workforce.

[61] Security through obscurity, or STO, is the belief that any system can be secure provided no one outside of its development group is allowed to be aware of its internal mechanisms.

commercial supply chain, commercial industry, and commercial standards, the better its access to SOTA microelectronics capabilities will become.[62]

In light of this problem, the Chief Scientist of the Air Force in 2023 led an independent study with 27 experts to study DoD's trusted foundries and "quantifiable assurance" approaches to microelectronics security.[63] The study found that a new system, MQA, can be used in the future in concert with the trusted foundries system to meet DoD's mix of needs for classified, unclassified, and export-controlled microelectronics. RAMP and RAMP-C, cited above, represent steps in this direction. MQA is a proposed system to assure confidentiality, integrity, and availability of microelectronics—with access being the overarching requirement. Because of the costs and ever-advancing technology, DoD cannot maintain its own dedicated facilities and relies on the commercial supply chain to meet the great bulk of its needs. Because the commercial chip industry must meet very-high-quality and performance standards and already collects the needed data to enable this, MQA would be a system for independent, data-centric checks on commercial process.[64]

However, a DoD Inspector General report from May 2022 indicates that the transition from a trusted foundry model to an evidence-based assurance alternative approach is "behind schedule" and attributes these delays to "difficulties in developing and staffing new processes" (coronavirus disease 2019 [SARS-CoV-2]), and turnover of key personnel.[65] In practice, this means that DoD is now 3 years behind the Section 224 requirements of the FY 2020 NDAA, which stipulated that DoD was required to establish trusted supply chain and operational security standards by January 2021 and that DoD purchases of microelectronics meet these standards effective January 2023. (See Recommendations 5.8, 5.9, and 5.15.)

---

[62] Recognition of the importance of more closely aligning DoD standards to the extent possible with commercial standards was conveyed at the 2023 Microelectronics Reliability and Qualification Workshop. See C. Rink, DoD OUSD S&T, 2023, "Microelectronics Policy, Standards and Guidance," presentation, January 23, https://csrc.nist.gov/presentations/2023/microelectronics-policy-standards-and-guidance.

[63] Department of the Air Force, Office of the Chief Scientist and NDAA, 2023, "Mandated Independent Review of USD (R&E) Microelectronics Quantifiable Assurance Effort," August 3, https://www.af.mil/Portals/1/documents/2023SAF/MQA_Report.pdf.

[64] Regarding how a testing and auditing system can operate in a related area, see Dario Amodei, CEO Anthropic, "Testimony to the Senate Judiciary Committee, Subcommittee on Privacy, Technology and the Law," July 25, 2023, https://www.judiciary.senate.gov/imo/media/doc/2023-07-26_-_testimony_-_amodei.pdf.

[65] DoD Inspector General, 2022, "Evaluation of the Department of Defense's Transition from a Trusted Foundry Model to a Quantifiable Assurance Method for Procuring Custom Microelectronics (DODIG-2022-084)."

IMMIGRATION

In addition, as will be discussed further in Chapter 7 on workforce education, the immigration system limits the ability to access and retain outstanding foreign science and engineering talent trained in semiconductor fields. As noted above, a recent study showed 40 percent of high-skilled semiconductor workers in the United States were born abroad. This situation calls out for improvement.[66] Overall, legislation is needed to revise the immigration laws to reflect the reality of technology needs, including that country of origin should not matter when immigrants are selected based on their skills and education, and to enable an accelerated path to citizenship. Meanwhile, without such comprehensive reform, an intermediate step is to award lawful permanent resident status for those with advanced science, technology, engineering, and mathematics (STEM) degrees under current law outside of country or worldwide numerical caps for individuals working in semiconductor fields (and other advanced technology areas). DoD could prioritize such approaches. (See Recommendation 7.2.)

OVERCOMING BARRIERS TO MODERNIZATION

The time it takes for DoD to field new platforms is a significant and growing problem[67]—a new weapons system can take a decade and a half or more to develop. This means that the platforms' information systems must be updated during development, increasing cost and delay problems. A new ship, aircraft, or tank will often operate for decades, which means its information systems require periodic updating to avoid becoming obsolete. As new electronics technologies evolve, they need to be embedded in existing platforms; yet Congress, in providing defense funding, has historically favored new procurements over modernization. The Defense Innovation Unit (DIU) was formed in 2016 to try to improve the time required for fielding AI, machine learning, autonomy, cyber, and energy savings technologies, largely through trying to rapidly leverage commercial technologies.[68] Cutting-edge advances in design, as recommended above, could be a significant enabler of DoD semiconductor modernization.

There is another consideration, as well. If China floods the market with legacy semiconductor chips, which it is now producing in growing volumes, then DoD faces a significant set of security issues because of its reliance on legacy chips in

---

[66] PCAST, 2022, "Report to the President on Revitalizing the Semiconductor Ecosystem," September.

[67] A. Seraphin, 2023, "Defense Department Semiconductor Concerns," Presentation to National Defense Industries Association, September 12.

[68] D. Vergun, 2023, "DoD Modernization Relies on Rapidly Leveraging Commercial Technology," *DoD News*, January 25, https://www.defense.gov/News/News-Stories/Article/Article/3277453/dod-modernization-relies-on-rapidly-leveraging-commercial-technology.

its many systems, providing an additional reason for reconsidering DoD's current modernization approaches. While DoD has a software modernization strategy,[69] a corresponding strategy for semiconductor hardware updates to improve system performance is needed. (See Recommendation 5.16.)

OVERCOMING WORKFORCE EDUCATION BARRIERS

There are problems noted above in the education systems affecting semiconductors at professional (engineers and scientists) and technical workforce levels, and DoD has long had programs in both areas. The National Science Foundation (NSF) and DOC have responsibilities under the CHIPS Act to form new workforce programs to address semiconductor needs. DoD needs to collaborate with NSF and DOC on these efforts and bring its own workforce education programs to bear. Chapter 7 of this report provides a detailed summary of workforce issues both for the semiconductor industry and DoD and corresponding recommendations.

All of the steps listed above could be considered by DoD as part of a coherent, DoD-wide general semiconductor strategy that it will need to develop to meet its semiconductor needs. Specific recommendations for the barriers discussed in this chapter are referenced above and contained in subsequent chapters. But the need for an overall DoD strategy and unified approach remains.

---

[69] DoD, 2021, "Department of Defense Software Modernization Strategy, Version 1.0," November, https://media.defense.gov/2022/Feb/03/2002932833/-1/-1/1/DEPARTMENT-OF-DEFENSE-SOFTWARE-MODERNIZATION-STRATEGY.PDF.

4

The Role of Public–Private Partnerships in Supporting Semiconductor Manufacturing

The committee was asked to review and evaluate how public–private partnerships (PPPs) can strengthen semiconductor manufacturing, explore how PPPs can be tailored to address different aspects of the semiconductor supply chain, and explore the merits of establishing a semiconductor manufacturing corporation.

The committee finds that PPPs have a mixed record in strengthening semiconductor manufacturing. There has been substantial research looking into this question, some of which has been funded by the Department of Defense (DoD), as discussed below. Taken together, this research suggests that PPPs can accelerate innovation in low technology readiness level (TRL) basic research that supports semiconductor manufacturing and facilitate mid-TRL applied research that accelerates technology development with an eye toward transition to product manufacturing. To succeed, PPPs must be substantive and have a well-articulated research agenda and goals, robust participation from industry, strong intellectual property (IP) protections, and consistent funding for multiyear periods. PPPs that lack one or more of these characteristics frequently fail to achieve their aims. It is not possible to create a one-size-fits-all PPP that can address the myriad economic and technical challenges facing the semiconductor industry, or address challenges in multiple parts of the semiconductor supply chain simultaneously. In general, PPP objectives must be clear, their scope must be manageable, and their participants and staff constantly engaged and supported. PPPs also need upstream and down stream interactions, as well as to know where they fit in the strategic vision and how to contribute to it.

This research leads the committee to conclude that a PPP can strengthen semiconductor manufacturing provided it has the following characteristics:

- It is well funded.
- It is funded at stable levels for 5–10 years, with annual reviews.
- It is focused on one TRL range, for example TRL 4–6 (the so-called "valley of death").
- It is focused on a specific supply chain sub-segment.
- It is structured to maximize the possibility of technology transfer to industry.
- It is not restricted solely to U.S. entity membership.
- Its research and IP (and eventually, product sales) can be shared internationally.

The committee does not find that establishment of a semiconductor manufacturing corporation that is a government-owned or government-operated captive fabrication facility (fab) and that exists solely to support DoD microelectronics needs is advisable. For the techno-economic reasons discussed in previous chapters, a DoD-centric semiconductor manufacturing corporation is highly unlikely to deliver the variety of mature and leading-edge semiconductors DoD requires at commercially competitive timelines and prices.

On the other hand, a public–private semiconductor manufacturing corporation that serves primarily as the administrative coordinator of DoD, other government, and private subsidies for commercially oriented, next-generation fabrication technology development and capacity expansion *might* be a concept worth exploring. That model would be analogous to the highly successful Semiconductor Research Corporation (SRC) public–private semiconductor research and development (R&D) partnership, but focused on nearer-term semiconductor design, fabrication, and deployment for technology nodes beyond those currently deployed.

With or without such a PPP, DoD will be well served by making every effort to ensure that leading-edge chip manufacturers are successful in increasing U.S.-based production that is commercially successful and viable and by being an engaged partner with these firms. This section focuses on how PPPs can support semiconductor manufacturing more generally and potentially address DoD-specific supply chain challenges.

DoD commissioned studies analyzing the role of PPPs in the microelectronics industry in 2017,[1] 2021,[2] and now 2023 (for this committee's work). These studies emphasize shared characteristics and themes of successful PPPs, which can inform

---

[1] Potomac Institute for Policy Studies, 2017, "Consortia Analysis and Recommendations Trade Study," https://www.potomacinstitute.org/reports/43-pips-reports/191-consortia-analysis-and-recommendations-trade-study.

[2] V. Pena, M.M.G. Slusarczuk, J. Mandelbaum, et al., 2021, "Lessons Learned from Public-Private Partnerships (PPPs) and Options to Establish a New Microelectronics PPP," IDA Document: D-22782, https://apps.dtic.mil/sti/trecms/pdf/AD1200224.pdf.

a model PPP to address one or more aspects of DoD's microelectronics challenges. Before proceeding with findings and recommendations in response to the second item in the statement of task ("How can public–private partnerships strengthen semiconductor manufacturing?"), the findings of these reports merit review.

POTOMAC INSTITUTE: CONSORTIA ANALYSIS AND RECOMMENDATIONS TRADE STUDY

In 2017, the Potomac Institute for Policy Studies conducted a 6-month study to identify lessons learned from the history of semiconductor R&D consortia and government R&D efforts involving technology transition, and to apply these to the current needs of the semiconductor industry. The study examined six different semiconductor PPPs, with a focus on identifying organizations in the mid-range TRLs (4–7). Case studies were developed on the Very High-Speed Integrated Circuits Program (VHSIC), the Semiconductor Manufacturing Technology Consortium (SEMATECH), Manufacturing USA, the Industrial Technology Research Institute (ITRI, in Taiwan), the Interuniversity Microelectronics Centre (IMEC, in Belgium), and the Faraday Centers (in the United Kingdom). This study made five recommendations based on its findings:

- *Funding:* Ensure that the funding level of any major public–private R&D initiative is at a magnitude that mirrors previous successful initiatives: $200–$300 million per year, 10-year timeline, total budget of $2 billion–$3 billion, in 2017 figures.
- *Technology transition:* Include costs for technology transition and insertion in the budget.
- *Major industry role:* Follow industry's lead while making sure the U.S. government still benefits. Ensure cost sharing to the extent feasible between government and industry.
- *Focus on low volume:* Focus public–private R&D efforts on low-volume, customizable manufacturing solutions to technical challenges.
- *Procurement vehicles:* Use Other Transaction Authority (OTA) acquisition mechanisms in all R&D programs.

INSTITUTE FOR DEFENSE ANALYSES: LESSONS LEARNED FROM PUBLIC–PRIVATE PARTNERSHIPS AND OPTIONS TO ESTABLISH A NEW MICROELECTRONICS PUBLIC–PRIVATE PARTNERSHIP

In 2021, the Defense Advanced Research Projects Agency (DARPA) funded the Institute for Defense Analyses to study and analyze prior and ongoing PPPs to inform objectives for a new PPP that could advance microelectronics R&D, potentially by supporting infrastructure to meet prototyping needs across the industry. This study surveyed a variety of PPPs, assessing their ability (or lack thereof) to achieve their stated goals. This study identified 32 lessons learned, several of which are particularly relevant to the exploration of how PPPs can strengthen semiconductor manufacturing.

- *Goals:* The goals of the PPP must be clearly defined and different visions on topics ranging from basic research, proof of concept testing, prototyping, and workforce development must be reconciled.
- *Governance:* A critical decision is the choice of governance model. Models observed ranged from consultative decision-making by executive leaders to consensus voting by members. The critical factor for success is that the governance is transparent and has the broad confidence and support of the PPP members. The realm of governance must include the high-level business strategy, the technical agenda and priorities, member engagement, and clearly defined success measures, and it must be consistent with the members' authorities and business models.
- *Funding:* The funding structure must be flexible to accommodate varied members' expectations and sufficient to support the PPP's mission. Funding structures tied to governance must permit resolution of different opinions to ensure that members do not exit the PPP and allow for a process to attract new partners as the PPP evolves.
- *Intellectual property policies:* IP policies and rules must be defined up front, and one should anticipate that this will be challenging, given the widely different approaches that exist in industry, university, and government and the need to harmonize them. These policies must adapt to a range of pre-existing commitments that prospective members have, their general policy approaches, and their attitudes toward physical security and export control. ITAR control of defense articles presents a significant obstacle for non-U.S. participation.
- *Evaluation and measures of success:* Measures of success must be tied to the PPP's goals. These measures may include technical milestones as well as economic and social returns, such as the creation of new businesses, jobs, and social well-being. Financial sustainability, or the degree of self-sufficiency from federal funds, is another success measure. However, the degree and timelines for self-sufficiency may vary depending on the PPP's goals and the scale of investments, for instance, in new R&D and prototyping infrastructure. The long-term support of the needed operational infrastructure should be considered as part of the PPP's funding model. In addition, for any PPP that operates on long timelines, ensuring regular evaluations of progress (both quarterly and annually) will inform government and industry participants eager to evaluate the pace and likelihood of breakthroughs.

This committee's findings align with the findings of these earlier studies. PPPs can strengthen semiconductor manufacturing and support DoD's needs, directly and indirectly. PPPs can directly strengthen semiconductor manufacturing by convening stakeholders (generally a mix of industry, academia, and government) and enabling and accelerating innovation (TRL 1–3), maturing existing technologies through prototyping (TRL 4–6), or serving as a vehicle to transfer and transition technologies to production (TRL 7–9).[3] PPPs can also indirectly strengthen semiconductor manufacturing by serving as a forum to engage unconventional parts of the semiconductor ecosystem, including venture capital, 2- and 4-year colleges and universities, start-ups, and local governments. Engagement with these otherwise disparate parts of the semiconductor ecosystem under the auspices of a PPP can result in knowledge transfer, technology transfer, and workforce training that ultimately supports semiconductor manufacturing.

This chapter proceeds by describing ways that PPPs can directly support semiconductor manufacturing (with examples), ways that PPPs can indirectly support semiconductor manufacturing, and key features of successful PPPs. Related to this chapter, there is a "playbook" in Appendix C that could be followed to establish a PPP that supports semiconductor manufacturing.

PUBLIC–PRIVATE PARTNERSHIPS AND DIRECT SUPPORT TO SEMICONDUCTOR MANUFACTURING

Recalling the findings of earlier studies[4] on semiconductor PPPs that emphasized the need for clear goals, narrow scoping, and measurable metrics of success, in general, these PPPs should only focus on *one* range of the lines of effort: TRL 1–3, TRL 4–6, or TRL 7–9. For example, a PPP that attempts to investigate the materials science behind a prospective post-complementary metal-oxide-semiconductor

---

[3] Technology readiness levels (TRLs) are a type of measurement system used to assess the maturity level of a particular technology. Each technology project is evaluated against the parameters for each technology level and is then assigned a TRL rating based on the project's progress. There are nine technology readiness levels. TRL 1 is the lowest, and TRL 9 is the highest. For example: TRL 1—When a technology is at TRL 1, scientific research is beginning and those results are being translated into future research and development. TRL 4—Once a proof-of-concept technology is ready, the technology advances to TRL 4. During TRL 4, multiple component pieces are tested with one another. TRL 7—TRL 7 technology requires that the working model or prototype be demonstrated in a relevant environment (in the case of the semiconductor industry, this would be a fab). Also see https://www.nasa.gov/directorates/somd/space-communications-navigation-program/technology-readiness-levels.

[4] Potomac Institute, 2017, "Consortia Analysis and Recommendations Trade Study (CARTS)," Potomac Institute for Policy Studies, https://usmicroelectronics.mit.edu/wp-content/uploads/2022/06/DARPA-Consortia-analysis-and-recommendations.pdf; Pena et al., 2021, "Lessons Learned from Public-Private Partnerships (PPPs) and Options to Establish a New Microelectronics PPP."

(post-CMOS) technology while simultaneously developing the prototyping infrastructure to support volume manufacturing of that technology is at high risk of failing to succeed in either endeavor.

Given the scale and expense required to innovate and compete in the semiconductor industry, PPPs must be well-scoped to make progress on a discrete challenge and execute its research agenda in a manner that lends itself to technology transition to industry. Semiconductor PPPs engaged in research that lacks industry interest or commercial appeal will almost certainly fail to generate IP or innovations that meaningfully support semiconductor manufacturing. In general, PPPs can support semiconductor manufacturing by convening a consortium of members to make progress on a research agenda in one or more of the following areas.

Semiconductor Basic Research and Development (TRL 1–3)

PPPs can facilitate high-risk and high-reward foundational R&D into new materials, software tools, and instruction set architectures to establish proof-of-concept applications for semiconductor manufacturing. This research uses analytical and laboratory studies to establish the viability of a speculative technology before it is ready to proceed to the next stage of development. Often, a proof-of-concept may be generated at this stage.[5] However, it also usually necessary for basic research teams to have a sustained engagement throughout the product development process and the production process, since new scientific questions and technical challenges frequently arise at each stage.

As discussed in Chapter 3, there are a wide variety of post-CMOS materials and devices being explored by the semiconductor industry. A PPP could be established with a research agenda focused on investigating one or more of these post-CMOS materials for DoD and industry microelectronics uses. For example, a PPP focused on basic R&D could explore the suitability of carbon nanotube transistors or graphene field effect transistors as a post-CMOS alternative.[6,7]

Semiconductor Applied Research and Development (TRL 4–6)

PPPs can facilitate applied R&D that takes existing proofs-of-concept and validates their utility in a laboratory environment. At this stage, once the basic science of a technology has been established and a proof-of-concept generated, multiple

---

[5] C.G. Manning, 2023, "Technology Readiness Levels," NASA, September 27, https://www.nasa.gov/directorates/somd/space-communications-navigation-program/technology-readiness-levels.

[6] B. Ham, 2020. "Carbon Nanotube Transistors Make the Leap from Lab to Factory Floor," *MIT News*, June 1, https://news.mit.edu/2020/carbon-nanotube-transistors-factory-0601.

[7] A.P. Singh, P.N. Shankar, R. Baghel, and S. Tirkey, 2023, "A Review on Graphene Transistors," *2023 IEEE International Students' Conference on Electrical, Electronics and Computer Science (SCEECS)*, https://doi.org/10.1109/SCEECS57921.2023.10062965, pp. 1–6.

component pieces are tested with one another and simulations are run to validate that the technology functions in realistic environments (in this case, a fab). At this stage, a fully functional prototype may be developed.[8]

The Department of Energy (DOE) and three of its national laboratories, in partnership with Intel, Motorola, AMD, and Micron, established a 3-year PPP in the late 1990s to build a proof-of-concept extreme ultraviolet (EUV) lithography tool as part of the industry's efforts at the time to identify a next-generation lithography method.[9] This PPP was focused specifically on applied R&D: maturing the technology for use in a commercial high-volume semiconductor manufacturing fab. The result of this PPP was the so-called Engineering Test Stand, a functional prototype EUV tool and the predecessor to today's EUV tools sold by ASML.

Semiconductor Prototyping Infrastructure (TRL 7–9)

PPPs can create the physical infrastructure necessary for low-volume semiconductor manufacturing (also referred to as "nanofabs"), allowing faster feedback loops between creating a speculative new chip design and determining whether it works as intended. At this stage, prototypes are tested in a fab environment for incorporation into a high-volume semiconductor manufacturing flow and, if successful, the technology is incorporated into a manufacturing process.

For example, the Global450 Consortium (G450C) was established in 2012 (although efforts began as early as 2008) with the goal of developing the systems, process IP, and infrastructure necessary to aid the industry's then-expected transition from 300 mm wafers to 450 mm wafers (in general, the larger the wafer size, the more favorable the economies of scale).[10,11,12] G450C's membership consisted of SUNY Albany's College of Nanoscale Science and Engineering (CNSE), Intel, TSMC, Samsung, IBM, GlobalFoundries, and Nikon.[13] Although this consortium disbanded in 2017, and the industry has yet to adopt 450 mm wafer manufacturing, this is an example of a PPP that supported semiconductor manufacturing specifically through

---

[8] C.G. Manning, 2023, "Technology Readiness Levels," NASA, September 27, https://www.nasa.gov/directorates/somd/space-communications-navigation-program/technology-readiness-levels.

[9] D.D. Sweeney, 1999, "Extreme Ultraviolet Lithography," Lawrence Livermore National Laboratory, https://www.llnl.gov/sites/www/files/2020-05/euvl-STR-Nov-99.pdf.

[10] P. McLellan, 2022, "Whatever Happened to 450mm Wafers?" *Cadence*, August 18, https://community.cadence.com/cadence_blogs_8/b/breakfast-bytes/posts/450mm-wafers.

[11] Silicon Semiconductor, 2012, "Global Consortium to Make the Move to 450mm," August 21, https://siliconsemiconductor.net/article/75801/Global_Consortium_To_Make_The_Move_to_450mm.

[12] Intel, 2008, "Intel, Samsun Electronics, TSMC Reach Agreement for 450mm Wafer Manufacturing Transition: Companies Target Common Timeline for 450mm Wafer Pilot Line Readiness," https://www.intel.com/pressroom/archive/releases/2008/20080505corp.htm.

[13] J. Hruska, 2017, "450mm Silicon Wafers Aren't Happening Any Time Soon as Major Consortium Collapses," *ExtremeTech*, https://www.extremetech.com/computing/242699-450mm-silicon-wafers-arent-happening-time-soon-major-consortium-collapses.

development of novel prototyping infrastructure.[14] Semiconductor manufacturing customers validated the technical viability and economic appeal of 450 mm wafer manufacturing in fab environments, the relevant tooling and equipment innovations were in place, and infrastructure required to support the transition from 300 mm to 450 mm manufacturing was characterized.

Semiconductor Supply Chain

PPPs can pursue basic or applied research with the explicit goal of testing new materials, design tools, or equipment and resolve a known supply chain choke point. For example, the semiconductor industry's consumption of hydrofluorocarbons (HFCs), a greenhouse gas for which use is increasingly restricted internationally under the Kigali Amendment to the Montreal Protocol, necessitates that the industry develops one or more alternatives. HFCs are used by the semiconductor industry throughout the manufacturing process and are highly valued for their levels of purity. Ideally, an alternative will be developed by industry for industry, but there is a first mover problem in that no one firm has the resources to solve this challenge. A PPP could address this first mover problem and develop a standardized set of alternatives to HFCs to support semiconductor manufacturing through the formation of a consortia of members, innovating around this known choke point and limiting the industry's use of this environmentally hazardous material.[15,16,17]

PUBLIC–PRIVATE PARTNERSHIPS AND INDIRECT
SUPPORT TO SEMICONDUCTOR MANUFACTURING

PPPs can also strengthen semiconductor manufacturing indirectly, acting as a convening forum to accelerate innovation in each of the areas above and serve as a source of training and knowledge diffusion. By convening a broad

---

[14] S. Jones, 2022, "The Lost Opportunity for 450mm," SemiWiki [blog], April 15, https://semiwiki.com/semiconductor-services/techinsights/311026-the-lost-opportunity-for-450mm.

[15] Environmental Protection Agency, 2021, "Market Characterization of the U.S. Semiconductor Industry," https://www.epa.gov/sites/default/files/2021-03/documents/epa-hq-oar-2021-0044-0002_attachment_3-semiconductors.pdf.

[16] Semiconductor Industry Association (SIA), 2021, "Comments of the Semiconductor Industry Association (SIA) on the Notice of Proposed Rulemaking: Phasedown of Hydrofluorocarbons: Establishing the Allowance Allocation and Trading Program Under the American Innovation and Manufacturing Act," https://www.semiconductors.org/wp-content/uploads/2021/07/SIA-comments-on-EPA-HFC-Phasedown-Rule-july-2-2021.pdf.

[17] Department of State, 2023, "The Montreal Protocol on Substances That Deplete the Ozone Layer," https://www.state.gov/key-topics-office-of-environmental-quality-and-transboundary-issues/the-montreal-protocol-on-substances-that-deplete-the-ozone-layer.

membership that includes some of the entities listed below, PPPs can facilitate technology transfer to industry. PPPs with diverse membership and partnerships enjoy a variety of advantages, each of which indirectly can support semiconductor manufacturing.

- *Local and state governments:* Local and state governments have the ability to provide direct economic incentives (subsidies) as well as indirect incentives (regulatory facilitation and permitting) that can increase the speed and ease of PPP operations.
- *Start-ups and small businesses:* PPPs that explicitly include small and medium-sized businesses (in addition to established players) act as a convening forum whereby new firms can better understand their prospective customer or business partner priorities, increasing the changes of long-term success for the smaller entities. These small businesses and start-ups also gain exposure through their participation in PPPs insofar as larger established firms get to see novel approaches to industry challenges.
- *Venture capital:* PPPs can attract venture capital firms and ensure that innovations derived from the work of the PPP have a path to commercialization with private capital support (whether from corporate or commercial venture capital).
- *Colleges and universities:* PPPs can convene students, postgraduates, industry experts, and potentially federal laboratory employees in a forum that facilitates knowledge sharing and learning by doing.
- *Workforce:* PPPs, through their work with colleges and universities in particular, can serve to increase the number of trained workers that support the semiconductor industry. This opportunity is described in Chapter 7, where workforce issues more generally are discussed.
- *Information sharing:* PPPs can also support semiconductor manufacturing by serving as an information-sharing service. For example, PPPs can make the findings of their work entirely public, and by sharing important technical data broadly with industry, these findings can inform related research efforts that are happening elsewhere but not necessarily under the auspices of the PPP itself.

KEY FEATURES OF SUCCESSFUL PUBLIC–PRIVATE PARTNERSHIPS

Successful PPPs are easy to work with and for; have a clear value proposition that appeals to industry, academic, and government participants; and transparent operational structure that ensures all participants derive equitable benefit from the PPP's work. Successful PPPs that make genuine advances in semiconductor manufacturing must be funded with large and patient capital, given the long timelines necessary to innovate in the semiconductor industry. In practice, this means that a PPP needs to be clear about its research agenda from the start, have well-defined metrics for success, and clear IP provisions. A PPP with these features will appeal to the semiconductor industry, whose participation is essential. Importantly, participation from industry also may allow for easier and faster access to physical facilities that are not otherwise available.

This section describes each of these features in greater detail, concluding with examples of PPPs that focus on different scopes of work and the various features necessary to ensure their success. See Box 4-1 for a PPP case study.

Scope and Clarity of Goals

A semiconductor PPP cannot and should not endeavor to solve broad problems facing this industry (e.g., "invent the future"). While such efforts are necessary, they are more appropriately directed by entities with skill in managing complex R&D agendas (such as DARPA). In general, PPPs are not optimal for basic R&D work that supports semiconductor manufacturing. Instead, PPPs are instead better positioned to address TRL 4–6 problems, as recognized by DoD's 2023 Microelectronics Commons (ME Commons) effort. Cost sharing will always vary based on the level of risk in a PPP's scope. Thus, TRL 1–3 tasks indicate higher risk, and thus one expects higher government investment, whereas TRL 4–6 or 7–9 tasks indicate somewhat lower risk and more balanced investment between government and industry. The scale of resources required to address most of the current concerns in the semiconductor sector, from workforce to capital, is simply too large for any one PPP to make broad progress on an industry-relevant timeline. Rather, individual semiconductor PPPs should address discrete challenges (i.e., as demonstrated by earlier PPPs that sought to "mature EUV technology for use in a fab" and "develop a pre-competitive 450 mm wafer prototyping fab facility") with a clear vision of success and transition to industry for commercialization.

Well-scoped PPPs focus on a narrow band of TRL and a well-defined semiconductor supply chain sub-segment. For example, a PPP focused on enabling early-stage R&D (TRL 1–3) and prototyping (TRL 4–6) of, say, new electronic design automation (EDA) tools will necessarily have a different structure, goals, governance, funding, and participation than a PPP focused on maturing a specific semiconductor manufacturing process technology in preparation for its inclusion in high-volume manufacturing (TRL 7–9). Likewise, a PPP that aims to advance breakthroughs in upstream basic research in novel semiconducting materials (such as graphene or carbon nanotubes) is not well suited to simultaneously conduct downstream applied research on EDA tools.

BOX 4-1
Public–Private Partnership Case Study: Semiconductor Research Corporation

Founded in 1982, the Semiconductor Research Corporation (SRC) is a U.S.-based public–private partnership (PPP) that convenes government agencies, universities, and companies to advance precompetitive technologies through collaborative research, and to educate the technology leaders of tomorrow.[a] Since its launch, SRC reports that it has funded more than $2 billion in research, sponsored 12,000 graduate students, and developed more than 700 patents.[b] SRC's two most recent lines of effort, each of which may be considered a standalone PPP, commenced in January 2018, focused on workforce development and future microelectronics.

- *Workforce development.* Through the Joint University Microelectronics Program (JUMP) and JUMP 2.0, SRC convenes industry participants and universities with coordination from the Defense Advanced Research Projects Agency (DARPA) to support large academic research centers focused on microelectronics. According to SRC, JUMP supports "long-term research focused on high performance, energy efficient microelectronics for end-to-end sensing and actuation, signal and information processing, communication, computing, and storage solutions that are cost-effective and secure."[c] The 2021 Institute for Defense Analyses (IDA) semiconductor PPP study reported that JUMP was a $200 million program, with 40 percent of the funds coming from DARPA and the remainder being provided by industry.[d] SRC reports that since inception, JUMP has funded 35 universities (1,307 students and 169 faculty) as well as more than 4,000 publications and 67 patent applications.[e]
- *Nanoelectronic Computing Research (nCORE).* Through nCORE, SRC convenes industry and academia with support from the National Science Foundation (NSF) and the National Institute of Standards and Technology (NIST). The primary scope of nCORE is exploration of "fundamental materials, devices, and interconnect solutions to enable future computing and storage paradigms beyond conventional CMOS, beyond von Neumann architecture, or beyond classical information processing/storage."[f] The 2021 IDA semiconductor PPP study reported that nCORE was a $13 million program, with 60 percent of the funds coming in equal amounts from NSF and NIST and the remainder being provided by industry.[g] SRC reports that since inception nCORE has funded 30 universities (227 students, 91 faculty) as well as more than 1,000 publications and 29 patent applications.[h]

SRC, with its JUMP and nCORE programs, offers examples of successful semiconductor PPPs. Leading domestic and international semiconductor firms participate in each of the PPPs, and the research funded by SRC has supported the training of thousands of university researchers, developing a pipeline of talent to support the U.S. semiconductor sector.

---

[a] Semiconductor Research Corporation (SRC), 2024, "Programs," https://www.src.org/program.
[b] SRC, 2024, "About," https://www.src.org/about.
[c] SRC, 2024, "Joint University Microelectronics Program," https://www.src.org/program/jump.
[d] V. Peña, M.M.G. Slusarczuk, J. Mandelbaum, et al., 2021, "Lessons Learned from Public-Private Partnerships (PPPs) and Options to Establish a New Microelectronics PPP," IDA document D-22782, July, https://www.ida.org/research-and-publications/publications/all/l/le/lessons-learned-from-ppps-and-options-to-establish-a-new-microelectronics-ppp.
[e] Peña et al., 2021, p. M-20.
[f] SRC, 2024, "Nanoelectronic Computing Research," https://www.src.org/program/ncore.
[g] Peña et al., 2021, p. M-1.
[h] Peña et al., 2021, p. M-20.

Metrics for Success

In tandem with these clear goals and scope, a PPP should have a vision for success and metrics to evaluate progress. These metrics could consist of statistics, including (1) the total number of partners (industry and government agencies or entities), (2) the number of patents or copyrights filed and licensed within 10 years of program's start, and (3) the number of products, especially products acquired by the government that contain IP resulting from the collaborative work, as identified by performers and transition partners, or (4) the level of follow-on investment from non-government PPP members.

Intellectual Property

In general, IP treatment in a PPP should be determined collaboratively, ideally by an advisory council consisting of PPP participants who define mutually agreeable terms and conditions. There is no one-size-fits-all approach to IP treatment among semiconductor industry PPPs. Regardless of the particular part of the semiconductor supply chain being targeted by a given PPP, IP protections will have to follow one of three general options:

- *Public.* The results and findings from a PPP are shared with the public on an open-source or open-access basis.
- *PPP members only.* The results and findings from a PPP are shared with PPP members only. This model can offer members either an exclusive royalty-free license or a first-right-of-refusal opportunity to license the technology developed in the PPP.
- *PPP R&D performers only.* The most restrictive approach to PPP IP handling—the results and findings from a PPP are shared with only the subset of PPP members who perform the R&D.

In addition to these options, PPPs that make use of federal government R&D funds must also consider several important laws that afford the government special rights to the IP. These include the following:

- *Government purpose rights.* The government retains the right to use any breakthroughs a PPP may accomplish in associated government efforts.
- *March-in rights.* If the IP generated by a PPP is not being adequately commercialized, the U.S. government retains the right to make the patent(s) and license(s) available.

The treatment of IP in PPPs is addressed in greater detail in Chapter 5 in the section "Addressing Bureaucratic and Regulatory Barriers."

Timeline and Funding

PPPs need to be given adequate time and funding to succeed. PPPs need 6–12 months to solicit initial membership interest and funding. During this "phase 1" period, the scope of the work can be determined. In general, advances in TRL 1–3 or TRL 4–6 are possible within a period of 3–5 years. However, substantive advances in applied research, especially in semiconductor manufacturing and semiconductor manufacturing equipment, will likely require additional time.

PPPs require considerable funding to generate advances that will meaningfully address one or more of the challenges facing the semiconductor industry. The ratio of funds from government versus industry should depend on the level of risk associated with the PPPs research agenda: high-risk research requires greater government funding. Government funding is essential for semiconductor PPPs because it provides "patient" and consistent support to stabilize the PPP against the risk that one or more industry partners might cease funding for one reason or another. For example, IMEC reports that approximately 75 percent of its annual operations are funded by semiconductor industry participants, while 25 percent of its operating funds come from a variety of government funding sources.[18] The fraction of government funding was much larger in the early years after IMEC's founding, as appropriate for an entity that was working to establish a new model for global R&D.

Physical Facilities

A semiconductor PPP needs access to facilities and equipment, as well as technical and administrative staff. Ideally a preexisting facility can be leveraged (to maximize the speed with which a PPP begins operations), but importantly, this facility should be independent and not onsite at a single government or industry participant location. The larger of the PPP participants should be expected to contribute technical staff to this new facility, increasing their commitment, limiting freeriding, and maximizing engagement. As an alternative approach, participants may choose to develop technology in their respective facilities pursuant to an IP agreement to collaborate at the pre-patent and pre-commercial stage. This arrangement has the benefit of allowing each partner to control their IP development internally while still being obligated to share it with select partners. Importantly, however, this model only works well if the companies are not competitors (e.g., if participant companies support different parts of the supply chain).

---

[18] Potomac Institute, 2017, "Consortia Analysis and Recommendations Trade Study," Potomac Institute for Policy Studies, https://potomacinstitute.org/images/studies/CARTSsm.pdf.

Public–Private Partnership Membership and Industry Alignment

A semiconductor PPP should be led and funded primarily by industry; however, the specific levels of participation or funding vary commensurate with level of risk of the PPP's research agenda. Maximizing industry leadership, engagement, and participation is the single most important factor when ensuring a PPP supports a manufacturing objective. Crafting a structure that incentivizes industry to contribute technical staff to collaborate in a precompetitive forum is ideal.

Different PPP projects will inevitably have varying levels of industry participation depending on how short or long term the project is and whether the project involves basic or applied R&D. For example, a PPP focused on TRL 1–3 will likely see its research agenda reflect the priorities of universities and government laboratories that focus much of their work at this TRL level, while a PPP focused on TRL 7–9 will have a research agenda much more tailored to the commercial partners eager to transition the innovation to high-volume manufacturing. Also, industry involvement will be affected by the nature of the projects the PPP pursues. For example, as the 5-year timetable for the government funding for the Manufacturing USA Institutes ramped down, only large firms were able to provide support for continuing the institutes. However, these larger firms could not justify investments in projects that would not directly benefit them, such as for SMEs, or projects that trained workforces other than their own.

Well-scoped PPPs ensure that their membership and industry alignment reflects the type of R&D the organization will engage in, the consolidation present in that particular sub-segment of the supply chain, and the level of capital required for the PPP to realize its desired goals. For example, a semiconductor sector PPP that hopes to make meaningful innovations in the field of photoresists will likely have to include membership from prospective customers (fabrication plants, or "fabs"), equipment vendors that make use of this photoresist, established photoresist manufacturers, and academic experts in the field. While this PPP will require substantial capital, photoresist manufacturing advances do not necessitate access to a high-volume fab at the prototyping stage, meaning that capital costs for the PPP will be relatively low. In contrast, a PPP that is focused on facilitating advances in semiconductor lithography will require participation from established industry leaders (both suppliers and customers of current lithography tools) and the substantial capital required to make progress in this high-cost segment of the tooling supply chain.

For PPPs that focus on early-stage research and development or prototyping, PPP membership should consist of organizations that excel at basic R&D. In practice, basic R&D in the United States is most frequently carried out by federally funded research and development centers (FFRDCs) and academia. As such, a PPP focused on early-stage research would have membership that consists of representatives from these fields and governance that reflects those priorities. PPPs focused on early-stage R&D are also at the stage where the government can exert the most influence over the direction and outcomes of a PPP's research program. PPPs focused on early-stage research will involve smaller sums of money and a focus on high-risk, high-reward research that can be crafted to align with government priorities, including those of DoD. In contrast, PPPs focused on later-stage applied research and semiconductor manufacturing will involve much higher capital costs and strong industry representation with an eye toward commercialization.

For PPPs that are focused on later-stage applied R&D as well as technology maturation in anticipation of transition to industry, membership and governance will necessarily focus more on ensuring greater representation from industry. The semiconductor industry is mature and heavily consolidated, which necessitates that PPPs designed to target a particular semiconductor supply chain sub-segment must ensure representation from existing industry players. This representation will ensure that the PPPs' work is aligned with industry priorities (i.e., in line with concurrent research on the topic happening in standards-setting bodies or other precompetitive forums) as well as offering industry the opportunity to evaluate the PPPs' work first-hand and assess whether the IP being generated is suitable for license and company use.

Table 4-1 presents examples of how a PPP could be tailored, based on the features described in this section, to address specific semiconductor challenges. Table 4-1 illustrates examples of notional PPPs and does not constitute committee recommendations.

Finding 4.1: The reports summarized above make clear that PPPs with the following characteristics can support semiconductor manufacturing: well-articulated goals and scoping and a clear definition of "success"; support with stable levels of funding for extended periods of time; a governance model that is supported by all participants; substantive engagement and leadership from industry; the necessary staff, equipment, and facilities to conduct their R&D agendas; emphasis on low-volume and customizable manufacturing solutions that lend themselves to technology transition to industry for scaled-up manufacturing; flexible funding levels that allow for new entrants while retaining existing members; and clear and consensus-based IP policies.

Recommendation 4.1: The Department of Defense (DoD) Office of the Under Secretary of Defense for Research and Engineering should strengthen public–private partnerships (PPPs) that support semiconductor manufacturing, particularly for technology readiness levels 1–6. DoD should ensure these PPPs have long-range funding, agreed-upon intellectual property terms, clear goals, and success metrics.

TABLE 4-1 Examples of How a Public–Private Partnership (PPP) Could Strengthen Semiconductor Manufacturing and Address Different Aspects of the Supply Chain

	Semiconductor Supply Chain Segment	Example of How PPPs Can Strengthen Semiconductor Manufacturing	How a PPP Could Be Tailored to Address Different Aspects of the Supply Chain
Upstream supply chain	Raw materials	New materials discovery and/or qualification	*Scope*: TRL 1–3 *Membership and industry alignment*: Strong government and academia representation *Intellectual property*: Open *Timeline and funding*: $50 million per year for 5 years
	Electronic design automation	Integrating AI advances with EDA tools to accelerate chip design timelines	*Scope*: TRL 7–9 *Membership and industry alignment*: Strong industry representation *Intellectual property*: PPP R&D performers only *Timeline and funding*: $500 million per year for 10 years
	Fabless/design	Piloting a hardware/software co-design approach using open-source instruction set architectures (e.g., RISC-V)	*Scope*: TRL 4–6 *Membership and industry alignment*: Balanced government, industry, and academia *Intellectual property*: PPP R&D performers only *Timeline and funding*: $100 million per year for 5 years
Downstream supply chain	Semiconductor manufacturing equipment	Increasing capture, recovery, and reuse of gases used in deposition tools	*Scope*: TRL 4–6; TRL 7–9 *Membership and industry alignment*: Strong industry representation *Intellectual property*: PPP R&D performers only *Timeline and funding*: $100 million per year for 10 years
	Manufacturing capability	Establish a prototyping fab and equipment ecosystem to transition from 300 mm to 450 mm wafers	*Scope*: TRL 4–6; TRL 7–9 *Membership and industry alignment*: Strong industry representation *Intellectual property*: PPP R&D performers only *Timeline and funding*: $250 million per year for 10 years
	Manufacturing capacity and resilience	Shorten laboratory to fab transition timelines through advances in multi-project wafer runs	*Scope*: TRL 4–6; TRL 7–9 *Membership and industry alignment*: Strong industry representation *Intellectual property*: PPP members only *Timeline and funding*: $250 million per year for 10 years
	Assembly, testing, and packaging	Accelerating advances in the chiplet ecosystem and heterogeneous integration	*Scope*: TRL 1–3; TRL 4–6 *Membership and industry alignment*: Balanced government and industry representation *Intellectual property*: Open or PPP members only *Timeline and funding*: $100 million per year for 5 years

NOTE: AI, artificial intelligence; EDA, electronic design automation; fab, fabrication; R&D, research and development; TRL, technology readiness level.

In light of rapidly evolving chip advances to improve security, size, weight, and performance, it is also important that hardware updates be incorporated effectively and efficiently into older systems on a much more frequent and ongoing basis. Planned updates should be factored into the design process for new systems. DoD's engagement with PPPs should reflect these points, working in concert with private-sector partners to deploy modern EDA tools, including those that make use of AI, to facilitate and streamline the replacement of older chipsets.

Finding 4.2: DoD has unique challenges related to obsolescent microelectronics that are no longer produced by the commercial microelectronics industry. Reconstituting production of legacy microelectronics is occasionally necessary to support specific military systems and requirements. The cost to DoD for production of these obsolescent microelectronics is high, and rising, owing to the dated design libraries that were used when these microelectronics were originally manufactured. In some cases, the original design files may not even be in a digital form.

Recommendation 4.2: The Department of Defense (DoD) (through the Office of the Under Secretary of Defense for Acquisition and Sustainment), as part of DoD's ongoing or future public–private partnerships, should develop new processes and practices to accelerate the deployment of advanced chips into existing DoD platforms and equipment, to deploy modern electronic design automation tools, incorporating artificial intelligence, to facilitate and streamline the replacement of older chipsets, thereby addressing security risks in legacy chips while improving size, weight, and performance where possible.

TARGETING PUBLIC-PRIVATE PARTNERSHIPS TO ADDRESS SPECIFIC DEPARTMENT OF DEFENSE EQUITIES IN THE SEMICONDUCTOR SUPPLY CHAIN

The committee was asked to address whether and how a PPP could support semiconductor manufacturing and the associated supply chain, including tool manufacturing, fabless design, electronic design automation, software development, manufacturing capability and capacity, workforce development, domestic research and engineering capture (e.g., hardware start-ups), and raw materials (e.g., wafers, rare earths). This section responds to these prompts by following the manner in which the semiconductor supply chain is organized, beginning with raw materials and concluding with assembly, testing, and packaging. This section focuses on DoD-specific considerations related to each aspect of this supply chain.

In general, the semiconductor supply chain consists of R&D, design, manufacturing, and assembly, testing, and packaging. Within each of these broad segments there are many sub-segments. For example, within "manufacturing," the supply chain consists of raw and processed materials (e.g., bare wafers and high-purity process gases), photoresist, photomasks, semiconductor manufacturing equipment, and fab infrastructure. Each of these supply chain sub-segments requires different subject-matter expertise, from materials science and chemistry to microscopy and device physics. This section describes DoD equities in each of these areas of the supply chain, emphasizing opportunities where a PPP may increase semiconductor supply chain resilience.

Materials (Chemicals, Gases, and Substrates)

Raw and processed materials are consumed intensively throughout the semiconductor manufacturing process. Roughly two-thirds of all elements in the periodic table have been used in the semiconductor manufacturing process at one time or another.[19] The U.S. Geological Survey, which maintains a list of 35 critical minerals, identifies 30 of these as relevant to semiconductor manufacturing. These critical minerals are not uniquely critical to the semiconductor industry. For example, these minerals may also be critical for the aerospace or large-capacity battery industries.

In addition to these raw elements, specialty materials like high-purity isopropyl alcohol, electronic-grade gases, chemical mechanical planarization slurries, and physical vapor deposition targets are used in semiconductor manufacturing. As a result of this tremendous complexity, firms specialize in the supply of one or more of these sub-markets. Many of these materials are commoditized, meaning there is a relatively small profit margin and whatever profit can be derived from their supply is not sufficient to fund large R&D budgets. Furthermore, because much of the R&D for these commoditized materials is relatively easily copied, firms face a further disincentive to invest in materials innovation research. Adding to these challenges in materials R&D, total consolidated worldwide demand for some of these specialty suppliers only supports one to three firms. As a result, increasing overall supply through subsidies that incentivize U.S.-origin manufacturing is inadvisable and would result in a scenario where oversupply may harm firms and affect availability. For example, funding entirely new photoresist production in the United States could distort the market such that overall availability of photoresist is reduced as profits fall and firms exit the market in the medium to long term.

U.S. firms like DuPont, Brewer Scientific, Photronics, and Entegris are active (and in some cases lead) in select materials sub-markets, while other markets

---

[19] N. Draeger, 2019, "Happy 150th Birthday to the Periodic Table," LAM Research, December 5, https://newsroom.lamresearch.com/Happy-150th-Birthday-to-the-Periodic-Table.

are dominated by firms in Europe and the Asia-Pacific. Notably, Japanese firms maintain a strong position in the supply of photoresists and photomasks as well as the supply of bare wafers, where Taiwanese firms such as GlobalWafers also have significant market share.

DoD is exposed to, and relies on innovation in, the materials that support semiconductor manufacturing. Previous DoD reports have examined its reliance on rare-earth element supply chains as well as raw material supply chains.[20] DoD maintains organizations such as the Defense Logistics Agency whose purpose is to ensure a stable supply of national security–critical raw and processed materials.[21] In addition, DARPA recently convened a workshop focused on novel approaches to rare-earth element separation.[22] However, the Defense Logistics Agency and other DoD organizations are not tasked specifically with stockpiling specialty semiconductor materials and upstream products, nor reshoring production of these materials (e.g., through the Defense Production Act). Separately from DoD efforts, DOE is pursuing a variety of initiatives funded via the Bipartisan Infrastructure Law to increase domestic critical mineral and rare-earth element availability, which may have positive indirect effects for DoD supply chain resilience to the extent they reshore U.S. production.[23]

Finding 4.3: DoD has previously conducted studies into a wide variety of raw material choke points that may affect DoD systems and readiness, and also maintains components, such as the Defense Logistics Agency, which are tasked with ensuring a stockpile and stable supply of many of the critical raw materials. Rare-earth elements are critical for a wide variety of DoD systems in addition to semiconductor manufacturing. Quantifying DoD demand for rare-earth elements and partnering with the DOE may increase domestic rare-earth element availability that supports DoD systems generally, and semiconductor manufacturing specifically.

Finding 4.4: There are dozens of announced projects in the United States under way today to expand U.S.-based production of many specialty materials used in semiconductor manufacturing, from electronic grade chemicals to bare wafer manufacturing, with many more expected as the Department of Commerce

---

[20] DoD, 2022, "Securing Defense-Critical Supply Chains: An Action Plan Developed in Response to President Biden's Executive Order 14017," February, https://media.defense.gov/2022/Feb/24/2002944158/-1/-1/1/DOD-EO-14017-REPORT-SECURING-DEFENSE-CRITICAL-SUPPLY-CHAINS.PDF.

[21] Defense Logistics Agency (DLA), 2024, "DLA Strategic Materials," https://www.dla.mil/Strategic-Materials.

[22] DARPA, 2023, "DARPA Seeks Input on Novel Methods to Separate, Purify Rare Earth Elements," July 12, https://www.darpa.mil/news-events/2023-07-12.

[23] Department of Energy, 2024, "Rare Earth Security Activities: Office of Manufacturing and Energy Supply Chains," https://www.energy.gov/mesc/rare-earth-security-activities.

(DOC) begins to distribute CHIPS Act incentives. However, given the large number of raw and processed materials used in semiconductor manufacturing, it is not realistic to expect that U.S.-based production alone will ever meet U.S. demand. To ensure a geographically diverse and robust integrated circuit supplier base, the U.S. government will need to identify strong suppliers in friendly countries that can augment the U.S. supplier base.

Recommendation 4.3: The Department of Defense (led by the Defense Logistics Agency in partnership with the director of the CHIPS Coordination Cell) should coordinate with the Departments of Energy, Commerce, and State to identify choke-point materials for semiconductor manufacturing and partner with private industry to actively cultivate a robust, geographically and geopolitically diverse supply base for each.

Research and Development

R&D is essential for semiconductor industry leadership. U.S. semiconductor firms lead the world in semiconductor R&D, whether measured by overall firm R&D spending levels ($40 billion in 2019) or R&D spending as a percent of net sales (an average of 18.75 percent in 2023).[24] DoD receives roughly $95 billion annually to fund its R&D activities, some of which is directed specifically to semiconductor manufacturing purposes.[25]

Notably, DoD received $2 billion under the CHIPS Act to support DoD-specific microelectronics research. Separately, DARPA has funded the Electronics Resurgence Initiative (ERI, and now ERI 2.0) and maintains a Microsystems Technology Office (MTO) with a variety of programs focused on inventing the future of microelectronics under the auspices of the Under Secretary of Defense for Research and Engineering.[26,27] The Under Secretary of Defense for Acquisition and Sustainment also manages a wide variety of programs focused on microelectronics R&D.

The most visible example of DoD's focus on microelectronics R&D was the 2023 establishment of the eight ME Commons hubs. These hubs, which each received initial funding in the tens of millions of dollars, aim to advance a wide

---

[24] SIA Factbook 2023 (18) and CSET Semiconductor Supply Chain report (12).

[25] GAO, 2023, "Research and Development: DOD Benefited from Financial Flexibilities But Could Do More to Maximize Their Use," GAO-23-105822, June 29, https://www.gao.gov/products/gao-23-105822.

[26] OUSD R&E, 2024, "Microelectronics Commons," https://www.cto.mil/ct/microelectronics/commons.

[27] DARPA, 2023, "Electronics Resurgence Initiative 2.0," March 6, https://www.darpa.mil/work-with-us/electronics-resurgence-initiative.

variety of microelectronics research initiatives relevant to DoD needs.[28] This work is happening simultaneously and separately from the work being guided by the DOC and NIST to establish the National Semiconductor Technology Center (NSTC) as well as related initiatives focused on advanced packaging, metrology, and a Manufacturing USA institute.

Finding 4.5: DoD is funding a wide variety of R&D initiatives via the ME Commons. DOC is also funding a wide variety of R&D initiatives in the coming years. Absent an effort to coordinate and align these activities, the United States will not see maximum return on its CHIPS Act investment. Formal cross-agency representation on advisory committees overseeing the work and technical exchanges is likely to be useful to ensure strategic alignment of government investments.

Recommendation 4.4: The Department of Defense (through the director of the CHIPS Coordination Cell) should ensure that its semiconductor research and development efforts are closely integrated with the related agendas funded through the Departments of Commerce and Energy and the National Science Foundation.

Electronic Design Automation and Software Development

EDA tools are the backbone of semiconductor design. These tools are essential for creating semiconductor designs, assisting engineers with the task of layout and defect detection. U.S. firms Cadence and Synopsys are dominant in this market, as is Mentor Graphics (now owned by Siemens, Germany). Combined, these three firms account for roughly 75 percent of global market share in EDA tools. These EDA firms are increasingly making use of AI to accelerate design cycle timelines and reduce the overall cost associated with designing a chip.[29]

Like all other chip customers, DoD relies on EDA tools for the design of custom legacy and leading-edge microelectronics, and in some cases likely uses EDA tools to assist in the integration of commercial-off-the-shelf microelectronics into its systems. DoD's purchase of specific EDA tools is siloed by the programs it funds. Separate from its use of EDA tools, DoD has also funded substantial research into future applications of EDA tools. DARPA's Intelligence Design of Electronic

---

[28] C.T. Lopez, 2023, "DOD Names 8 Locations to Serve as New 'Microelectronics Commons' Hubs," *DOD News*, September 20, https://www.defense.gov/News/News-Stories/Article/Article/3532338/dod-names-8-locations-to-serve-as-new-microelectronics-commons-hubs.

[29] J. Sawicki, 2020, "EDA in the Era of AI," *Electronic Design*, January 8, https://www.electronicdesign.com/markets/automation/article/21120058/mentor-a-siemens-business-eda-in-the-era-of-ai.

Assets (IDEA) program is one example of recent DoD focus on this supply chain segment.[30]

> Finding 4.6: The cost of microelectronics design has risen rapidly in recent years, driven in large part by costs associated with EDA tools and layout times. These cost increases affect DoD access to affordable and reliable microelectronics. In response to this challenge, the microelectronics industry is rapidly integrating AI as a means of automating chip layout, accelerating overall chip design times, and reducing design costs.[31] DoD has access to a large and unique set of data that could be leveraged to train an AI algorithm optimized for DoD chip design. However, these data are frequently siloed by program or subject to terms and conditions that restrict its use in AI training and inference tasks, necessitating legal resolution.

> Finding 4.7: DoD funds development of both EDA tools and chip designs, but these are not necessarily shared between the contractors working on specific DoD projects. Consolidation of these products into a clearinghouse might allow for faster access to leading-edge design tools and chip designs, generating cost savings for DoD by eliminating duplicative and dormant EDA licenses. A pilot project could be led by the Defense Microelectronics Activity (DMEA) or NSWC-Crane or similar organization. (See Recommendations 5.5, 5.6, and 5.7.)

> **Recommendation 4.5: The Department of Defense (DoD) (through the Office of the Under Secretary of Defense for Acquisition and Sustainment) should launch an effort that crosses service branches and programs to facilitate the cost-efficient use of electronic design automation tool licenses and establish a clearinghouse for DoD custom chip designs to be reused.**

Fabless Design

The United States remains the worldwide leader in semiconductor design. Importantly, however, DoD has specific semiconductor design requirements that diverge from the overall focus of the commercial semiconductor industry. This includes custom (not commercial-off-the-shelf) radiation-hardened microelectronics capable of operating in space or nuclear environments. Such microelectronics form the backbone of satellite communications and data handling as well as sensors for warning, positioning, navigation, and timing related to everything from offensive to defensive capabilities.

---

[30] S.K. Lim, 2024, "Intelligent Desing of Electronic Assets (IDEA)," DARPA, https://www.darpa.mil/program/intelligent-design-of-electronic-assets.

[31] Synopsys, 2024, "DSO.ai: Achieve PPA Targets Faster with the World's First AI Application for Chip Design," https://www.synopsys.com/ai/ai-powered-eda/dso-ai.html.

Custom radiation-hardened microelectronics have limited commercial appeal. A 2022 report from Deloitte estimated that, of the $660 billion global microelectronics market, radiation-hardened microelectronics only accounted for $1.5 billion.[32] While used in commercial spaceflight and satellites, most radiation-hardened microelectronics are consumed by governments for military applications. A 2023 study from the Defense Science Board (DSB) found that "today's microelectronics market is dominated by consumer-demand, rather than defense-demand, and the base of RH [radiation-hardened] and SRH [strategic radiation–hardened] suppliers has been shrinking. This shrinkage compounded by the dramatic consolidation of SOTA microelectronics manufacturing capability comprises a profound national security risk."[33] As a result of low commercial demand, DoD access to custom mature and leading-edge, radiation-hardened microelectronics is limited: there are very few firms in the United States operating in this market, and those that do offer a relatively small selection of products. For example, within the 17 foundries that are a part of the Trusted Foundry Program, only four of them offer any radiation-hardened products and among those that do, the most advanced node is 90 nm (which was introduced roughly 20 years ago).[34,35]

Importantly, however, as the commercial industry has advanced and fabless designs transition from a fin field-effect transistor (FinFET) to gate-all-around FETs (GAAFETs) and (eventually) to complementary FETs, research has demonstrated that commercially available leading-edge microelectronics have some radiation tolerance by default. Simply put, as device features get smaller, there is less opportunity for radiation-induced device upsets. The same 2023 report from DSB describes this development in detail, noting,

> A possible solution for long-term design, manufacture, and accessibility to RH/SRH may reside in leveraging capabilities of SOTA (or near-SOTA) foundries, with appropriate understanding of the radiation effects at the device level and appropriate redesign of SOTA chips, as concluded in a recent JASON study. Current studies suggest that the reduced dimensions of SOTA microelectronics may render them more robust to "single event upsets" (SEUs).[36]

---

[32] D. Stewart, D. Jarvis, C. Simons, and G. Crossan, 2022, "That's Just Rad! Radiation-Hardened Chips Take Space Tech and Nuclear Energy to New Heights: The Next Generation of Rad-Hard Chips Is Helping Bring Devices Used in High-Radiation Environments into the 21st Century at Last," *Deloitte Insights*, November 30, https://www2.deloitte.com/us/en/insights/industry/technology/technology-media-and-telecom-predictions/2023/radiation-hardened-electronics-market.html.

[33] S.A. Jackson, 2023, "Balancing Openness and Security Across the DoD Academic Research Enterprise (ARE)," Defense Science Board, August 23, https://apps.dtic.mil/sti/citations/trecms/AD1212739, Appendix C.

[34] OUSD, 2024, "Accredited Suppliers," May 22, https://www.acq.osd.mil/asds/dmea/tapo/docs/tp/Accredited-Supplier-22May2024.pdf.

[35] M. LaPedus, 2002, "Intel Works Toward Ramp of 90-nm Process in 2003," *EE Times*, February 27, https://www.eetimes.com/intel-works-toward-ramp-of-90-nm-process-in-2003.

[36] DSB (Defense Science Board), 2022, "Radiation-Hard Microelectronics," JSR-21-05, https://dsb.cto.mil/reports/2020s/ARE_Final%20Report_08102023.pdf.

To better coordinate DoD radiation-hardened microelectronics equities, DoD has the Strategic Radiation-Hardened Electronics Council (SRHEC).[37] The SRHEC convenes representatives from across the Departments of Defense and Energy as well as the National Aeronautics and Space Administration (NASA) who manage programs that require radiation-hardened electronics. One major challenge identified by the SRHEC in recent years, in addition to domestic supply of radiation-hardened microelectronics, is domestic availability of the radiation sources and expertise needed to test microelectronics to determine their radiation hardness. The lack of infrastructure across the nation to test radiation-hardened electronics has led to greatly extended chip and system development timelines of as much as 7 years.[38] This challenge is echoed by an earlier National Academies of Sciences, Engineering, and Medicine report, which found that the U.S. domestic electronics radiation testing capacity was likely to continue to be oversubscribed to the increase in commercial satellites and a variety of DoD and DOE modernization programs.[39]

Finding 4.8: FinFET and now GAAFET may have inherent radiation hardness improvements over past planar technologies.[40] These properties are potentially valuable to DoD microelectronics community, but the opportunity to integrate these leading-edge microelectronics into DoD systems has not been fully characterized. The 2023 DSB report and the 2022 JASON study on radiation-hardened microelectronics encourage DoD to determine if commercial-off-the-shelf, leading-edge microelectronics have radiation tolerance levels suitable to support DoD missions. This objective might be accomplished through a pilot project partnership with leading-edge microelectronics suppliers such as Intel, TSMC, or Samsung to test and evaluate leading-edge microelectronics for radiation hardness. Doing so offers the potential to access leading-edge chips and lower costs.

Finding 4.9: U.S. domestic radiation testing capacity for electronics remains a concern for DoD as well as other U.S. government agencies with radiation

---

[37] J. Franco and J. Ross, 2021, "Strategic Radiation-Hardened (SRH) Electronics Council (SRHEC) Public Summary from Analysis of Alternatives (AoA) for Domestic Single-Event Effects (SEE) Test Facilities," Presented during VI. a. Virtual Networking Opportunity w AoA Briefing and Status session at the *NASA Electronic Parts and Packaging (NEPP) Program 2021 Domestic High-Energy Single-Event Effects (SEE) Testing Users Meeting*, April 13, https://nepp.nasa.gov/workshops/dhesee2021/presentations.cfm.

[38] Ibid.

[39] National Academies of Sciences, Engineering, and Medicine (NASEM), 2018, *Testing at the Speed of Light: The State of U.S. Electronic Parts Space Radiation Testing Infrastructure*, Washington, DC: The National Academies Press, https://doi.org/10.17226/24993.

[40] N. Pieper, Y. Xiong, A. Feeley, et al., 2022, "SRAM Multi-Cell Upset Vulnerability at the 5-nm FinFET Node," Department of Energy, July 1, https://www.osti.gov/servlets/purl/2004056.

hardness equities, notably DOE and NASA. Increasing radiation hardness testing infrastructure in the United States will support government missions as well as the commercial spaceflight industry. DOC and the SRHEC may be valuable partners to determine what specific capabilities are necessary.

Recommendation 4.6: The Department of Defense (through the Strategic Radiation-Hardened Electronics Council in coordination with the director of the CHIPS Coordination Cell) should partner with the Department of Commerce to establish a national center of excellence for radiation-hardened microelectronics design and testing to greatly accelerate the timeline for development of such components for national security and commercial applications.

This national center of excellence could be tasked with increasing the availability of radiation-hardened microelectronics in the United States. It could explore the suitability of expected innovations in GAAFETs and chiplet architectures to support DoD's radiation-hardened microelectronics needs. DoD could partner with DOC to determine if funds allocated through the CHIPS and Science Act of 2022 could be used to create the national center of excellence with the needed radiation hardness testing facilities and expertise to efficiently support ongoing demand from DoD, NASA, DOE, and the commercial spaceflight industry. DoD could ensure that this center of excellence engages with leading-edge microelectronics suppliers to evaluate their products' potential radiation hardness and suitability to meet DoD needs.

Tool Manufacturing

The process of manufacturing semiconductors requires hundreds of sequential steps, each of which in turn relies on specialty equipment (semiconductor manufacturing equipment, or just "tools"). DoD has historically supported R&D in a variety of semiconductor manufacturing equipment technologies. Notably, DARPA funded a 15-year effort into next-generation lithographic technologies from the 1990s into the mid-2000s.[41]

In recent years, however, semiconductor manufacturing equipment has become increasingly capital-intensive, and the number of suppliers worldwide has consolidated to the point where there are only one to three suppliers for most major categories of tools. As a result of the increased capital expenditures necessary to compete in semiconductor manufacturing equipment and the overall consolidation

---

[41] *EE Times*, 2005, "DARPA Ends Litho Aid at Critical Juncture for Maskless," February 28, https://www.eetimes.com/darpa-ends-litho-aid-at-critical-juncture-for-maskless.

in the industry, the role of the government in supporting innovation in semiconductor manufacturing equipment has declined.

The CHIPS Act funded a $2 billion effort led by NIST to support metrology innovation.[42] In semiconductor manufacturing, metrology tools are used to ensure that fabricated microelectronics perform as intended and, in the event they do not, assist with defect detection and mitigation. These tools are among the most specialized types of equipment used in semiconductor manufacturing, and the overall number of suppliers in the United States and worldwide is limited.[43,44] Innovation in metrology tools supports innovation in the semiconductor sector generally, including advanced packaging, chiplets, and heterogeneous integration. NIST also has recognized the importance of developing a secure ecosystem for sharing metrology data to accelerate R&D related to semiconductor technologies, with the launch of its Metrology Exchange to Innovate in Semiconductors (METIS) program.[45]

Finding 4.10: Semiconductor manufacturing equipment is expensive, and the industry is heavily consolidated. Firms in the United States lead in many submarkets for semiconductor manufacturing equipment, including deposition (e.g., Applied Materials), etch (e.g., Lam Research), and metrology (e.g., KLA). DOC, through NIST, is funding a new initiative on semiconductor metrology equipment and process innovation to push the leading-edge of technology in this sub-market and sustain U.S. leadership. DoD also needs these efforts to succeed to ensure national security needs are met.

Recommendation 4.7: The Department of Defense (through the director of the CHIPS Coordination Cell) should coordinate with the National Institute of Standards and Technology to invest in new metrology technologies that support next-generation semiconductor manufacturing.

---

[42] NIST, 2024, "Notice of Funding Opportunity (NOFO)," https://www.nist.gov/chips/research-development-programs/metrology-program.

[43] KLA (Keep Looking Ahead), 2019, "KLA-Tencor Corporation to Change Name to KLA Corporation to Represent the Company's Broader Scope and Optimistic Vision," January 10, https://ir.kla.com/news-events/press-releases/detail/52/kla-tencor-corporation-to-change-name-to-kla-corporation-to.

[44] D. Rijnberk, 2023, "KLA Corporation: A Top Pick in the Semiconductor Industry," Seeking Alpha, November 27, https://seekingalpha.com/article/4654256-kla-corporation-a-top-pick-in-the-semiconductor-industry.

[45] NIST, 2023, "Building a Metrology Exchange to Innovate in Semiconductors (METIS): A Vision for the CHIPS Metrology Program Data Exchange," NIST CHIPS 1000-2 ipd, December, https://nvlpubs.nist.gov/nistpubs/CHIPS/NIST.CHIPS.1000-2.ipd.pdf.

Manufacturing Capacity and Capability

Through the CHIPS Act, Congress appropriated $39 billion to provide incentives to firms in the semiconductor industry to increase domestic production of microelectronics and support the underlying supply chain of products and services that support semiconductor manufacturing. A wide variety of firms have announced projects to create new semiconductor manufacturing capacity and capabilities in the United States. Notably, Intel, Samsung, and TSMC have announced new logic chip production facilities, Micron has announced a new memory production facility, Amkor and SK Hynix have announced new advanced packaging facilities, and a wide variety of materials and equipment suppliers are increasing domestic capacity and capabilities.

DoD stands to benefit from forthcoming increasing domestic availability of manufacturing capacity and capability. Notably, DOC has announced agreements with BAE Systems and Microchip Technology to provide financial incentives that will assist these firms' existing supply of microelectronics to DoD. While these efforts are focused on supporting DoD suppliers of custom mature microelectronics, DoD also requires access to leading-edge microelectronics as well. As described in the first item of the committee's statement of task, DoD's access to leading-edge microelectronics will only be possible through partnerships with commercial firms that supply these microelectronic components.

Finding 4.11: DOC is in the process of providing financial incentives to firms to increase U.S. domestic semiconductor manufacturing capacity and capabilities. DoD requires a wide variety of microelectronics, including both custom and commercial-off-the-shelf chips, at both mature and leading-edge nodes. Recent announcements by DOC of incentives for DoD microelectronics suppliers will support overall DoD microelectronics supply chain resilience. DoD microelectronics supply chain resilience will also materially improve if leading-edge suppliers of microelectronics expand capacity and capabilities at their U.S. facilities.

Recommendation 4.8: The Department of Defense (DoD) (through the director of the CHIPS Coordination Cell) should closely coordinate with the Department of Commerce on CHIPS and Science Act of 2022 incentive decisions to ensure funds are directed toward firms that can meet DoD's needs.

Assembly, Testing, and Packaging

Assembly, testing, and packaging refers to the process of verifying the functionality of finished semiconductors, protecting them, and creating the structures for physically connecting them into the systems for which they are designed to be incorporated. The semiconductor industry is currently pursuing a variety of advanced packaging efforts as Moore's Law slows down. Advanced packaging offers increased performance, power, modularity, and durability relative to traditional packaging approaches.[46] All of these attributes are of value to DoD as well as to the commercial sector.

DoD is thus similarly funding efforts to explore and leverage the promise of advanced packaging architectures for microelectronics. Notably, DoD has created the State-of-the-Art Heterogeneous Integrated Packaging (SHIP) Program specifically focused on creating a pathway to integrate commercial state-of-the-art packaging technologies for DoD systems.[47] This partnership leverages packaging innovations from Intel and Qorvo, which provide microelectronics to BAE Systems for integration in DoD systems. For example, Qorvo is reportedly supplying microelectronics that "provide form, fit, and function replacement of a legacy chip and wire design, resulting in significant production cost savings, reduced footprint, and improved performance."[48]

Finding 4.12: The microelectronics industry is investing heavily in advanced packaging technologies as a means of optimizing size, weight, power, and overall performance. DoD is similarly pursuing advanced packaging and three-dimensional heterogeneous integration technologies, notably through its SHIP program and ERI 2.0 and Next-Generation Microelectronics Manufacturing (NGMM). DOC is also expected to invest heavily in advanced packaging, specifically through the NSTC and the National Advanced Packaging Manufacturing Program (NAPMP). DOC is also expected to establish a new line of effort related to semiconductor metrology.

Recommendation 4.9: The Department of Defense (through the Office of the Under Secretary of Defense for Research and Engineering) should continue support for State-of-the-Art Heterogeneous Integrated Packaging (SHIP), SHIP 2.0, and other advanced semiconductor packaging research and development programs and initiatives, collaborating with commercial manufacturers to develop customized processes or capabilities when needed.

---

[46] J. VerWey, 2022, "Re-Shoring Advanced Semiconductor Packaging: Innovation, Supply Chain Security, and U.S. Leadership in the Semiconductor Industry," CSET, June, https://cset.georgetown.edu/publication/re-shoring-advanced-semiconductor-packaging.

[47] DoD, 2023, "Department of Defense Celebrates Advancements in Microelectronics Packaging Capabilities," April 6, https://www.defense.gov/News/Releases/Release/Article/3355049/department-of-defense-celebrates-advancements-in-microelectronics-packaging-cap.

[48] Ibid.

5

Challenges for the Department of Defense in Supporting Sustainability in the Semiconductor Ecosystem

BACKGROUND AND CONTEXT

The Department of Defense (DoD) has distinct electronics needs, particularly when it comes to electronics closely related to the physical layer, the portion of the system closest to the physical world such as close to an antenna or sensor. In domains such as optics, DoD uses different bands and frequencies than typical commercial applications. For applications such as radar, communications, and electronic warfare systems, DoD operates with unique frequencies and power levels. Meeting these niche requirements has required the development of new materials and manufacturing processes for semiconductors tailored to these specific needs.

DoD has been successful, to a significant extent, in fostering a community of firms that can cater to a series of these unique demands. In the realm of compound semiconductors, for example, numerous options are available, typically based on a 6-inch wafer production line. DoD has been able to support and cultivate this manufacturing at a cost and scale that aligns with the needs of these firms, ensuring they possess distinctive capabilities. Over the past few decades, this has been at the core of enhancing warfighting capabilities.

The United States has repeatedly pioneered new materials and integrated them into semiconductor production, leading to technologically advanced warfighting capabilities. These advancements have allowed DoD to detect adversaries before being detected, extend the reach and range of satellites, and gain a superior understanding of the battlefield compared with their adversaries. Work on integration of "exotic" materials typically takes a decade from invention to end-state manufacturing, but the investment has proven worthwhile in enabling new capabilities. In these activities, DoD has demonstrated that it can establish and sustain a thriving ecosystem. However, as systems have become increasingly digital in nature, the portion of such systems where DoD maintains this advantage has become progressively smaller.

As discussed in Chapter 3, DoD currently faces two challenges. The first is that it primarily operates in a low-volume, high-mix electronics environment, which is quite different from the high-scale production seen in consumer electronics. Consumer electronics thrive on producing millions of components, sometimes costing up to a billion dollars per chip development cycle. To make this investment worthwhile, the market opportunity for these chips usually needs to be very large. As technology has advanced, this scale requirement has become increasingly at odds with DoD's role in the electronics community. Although this issue is not unique to DoD, it has been particularly pronounced over the past two decades. Efforts to aggregate demand and change this dynamic have largely been unsuccessful. Even increasing purchases from 10,000 to 100,000 units falls short of the scale needed for modern development cycles, and unique requirements often remain unmet. DoD needs to accept and work within this low-volume, high-mix context. DoD has struggled to keep up with the complexities and costs of designing and manufacturing advanced technologies, where scalability is key (Box 5-1). These challenges stem from the nature of low-volume, high-mix markets and are not owing to poor decision-making or lack of expertise. They are common in both commercial and military sectors, which have different scalability factors.

A second challenge is DoD's diminished leadership in electronics research and development (R&D). In the past, DoD was closely involved in foundational R&D work such as finding the next generation transistor, such as the fin field-effect transistor (FinFET), but more recently it has intentionally focused on security or other more niche, DoD-specific programs, rather than driving the research agenda for the semiconductor sector. Nowadays, commercial factors drive modern semiconductor R&D, which is likely why the Department of Commerce (DOC) has been tasked with creating robust manufacturing operations in the United States through the CHIPS and Science Act of 2022 (CHIPS Act). DoD now needs to adjust to utilizing commercially focused manufacturing facilities without imposing barriers that could hinder U.S. manufacturing competitiveness on a global scale.

There is some alignment between the needs of DoD and the commercial sector in processing and storing large volumes of data, offering potential collaboration opportunities. However, certain policy decisions have complicated DoD's situation. For example, the Trusted Foundry program, designed to produce microelectronic components while maintaining security, has limited DoD's partnership opportunities. The Trusted Foundry program's manufacturing locations do not align with

---

BOX 5-1
The Unique Department of Defense Challenge for Artificial Intelligence

In the global competition surrounding the use of artificial intelligence (AI), the entity that controls the AI technology stack and supply chain will have a significant advantage in future conflicts. China has been rapidly advancing in developing a native AI technology stack. Recent demonstrations have revealed China's ability to create cutting-edge semiconductor electronics, surpassing the 10 nm node and producing graphic processing units (GPUs), the specialized chips for AI applications, that rival Nvidia's Tensor Core A100 class of GPUs for AI, currently a leading AI chip.[a]

Currently, a significant fraction of the value chain for advanced AI applications originates from U.S.-based companies. This includes the models, cloud infrastructure, accelerators, and electronic design automation (EDA) tools, and most chip manufacturing tools. This advantageous position is acknowledged worldwide, but the Department of Defense (DoD) has not capitalized on this inherent advantage to realize the potential of resulting defense applications.

The Defense Advanced Research Projects Agency (DARPA) recognized the impending AI revolution in 2017 through the AINext program, but instead of spearheading the revolution, DoD chose to study the technology and invest in various small initiatives, which did not lead to an AI infrastructure for defense needs. In contrast, the commercial industry transitioned from studying AI capabilities to productizing them in 2018. For instance, Microsoft's announcement of an AI supercomputer in 2020 can be traced to decisions made in 2018. This served as the foundation for OpenAI's ChatGPT capability, which launched in 2023.

Notably, DoD did not develop similar capabilities. To have developed an AI supercomputer, DoD would have needed to start around 2018 when commercial entities were pouring billions of dollars into infrastructure. However, owing to its own regulatory constraints around security requirements, creating its own chips would have required using the 90 nm node, which was the most advanced node supported by the Trusted Foundry program at the time. This is a concern, especially in light of recent announcements about Chinese GPUs created by Huawei.[b] China's dual alignment of industry with government gives it both commercial scale and military impact.

Fortunately, the U.S. industry maintains a substantial lead in AI, particularly in areas not directly tied to the unique physical layer. Here, the alignment of DoD needs and commercial interests largely overlaps. DoD requires access to AI technologies that are available from the commercial sector while ensuring a reliable supply during times of need.

---

[a] H. Mujtaba, 2023, "Chinese AI Company Claims Huawei AI GPUs Are on Par with NVIDIA A100, Will Compete with GPT-4 LLM in 2024," August 27, https://wccftech.com/chinese-ai-company-claims-huawei-ai-gpus-on-par-nvidia-a100-compete-gpt-4-2024/amp.

[b] AI Beat, 2024, "Huawei's Proprietary AI Chip Could Exceed Nvidia's A100 GPU," March 4.

those used by the advanced semiconductor industry, restricting DoD's access to cutting-edge chips. Furthermore, the sites within the Trusted Foundry program are now far behind the leading edge in providing modern fabrication options, further isolating DoD from industry trends, modern intellectual property, and advanced semiconductor manufacturing.

Beyond manufacturing challenges, designing modern systems is highly complex and resource intensive. Additional bureaucratic layers and unique steps in the design and manufacturing processes further complicate these efforts. Even with access to free manufacturing in a modern facility, DoD would struggle to create competitive designs using the most modern nodes. Similar issues affect the commercial industry, as designing at scaled nodes is challenging even for the largest entities, creating widespread competitive pressure in the industry.

ADDRESSING RESEARCH AND DEVELOPMENT BARRIERS

Chapter 3 discusses the challenges faced by the semiconductor industry as it approaches the physical limitations of current complementary metal-oxide-semiconductor (CMOS) technology, particularly as node sizes shrink to 3 nm and 2 nm. This situation indicates that the coming shift to post-CMOS technologies might be necessary within the next decade. To address these challenges, the community will need new architectures, materials, packaging, software, algorithms, and other innovative technologies.

Three primary lines of effort are particularly important: developing more efficient architectures and packaging, creating new models of computation, and discovering new materials and devices. The IEEE's International Roadmap for Devices and Systems (IRDS) has identified several beyond-CMOS approaches, including computational state variability, nonthermal equilibrium systems, novel energy transfer interactions, nanoscale thermal management, sublithographic management, and alternative architectures. Some promising advances are noted in the equipment sector with atomic layer deposition and etching, and in architectural developments such as in-memory processing, Nano/Micro Electro-Mechanical Systems (N/MEMS), magneto-electric spin-orbit (MESO) logic devices, cryogenic electronics, structures based on magnetic tunnel junctions, and edge computing. Additionally, three-dimensional (3D) heterogeneous integration packaging, which stacks multiple integrated circuits and connects them vertically for improved performance, lower power consumption, and reduced size, is currently receiving significant attention.

DoD is particularly interested in these advancements owing to their importance for national security as well as for economic security. DoD could play a pivotal role in coordinating and supporting next-generation R&D efforts in this domain. This is especially relevant as the funding for the proposed National Semiconductor Technology Center (NSTC) will conclude after 5 years. During this period, DoD should collaborate closely with the National Institute of Standards and Technology (NIST) on NSTC efforts and consider strategies for post-CHIPS Act funding.

An ongoing focus on post-CMOS R&D seems essential and DoD should fully partner on this work with NSTC in the near-term, while also developing its own long-term focus through DARPA's Electronics Resurgence Initiative (ERI) 2.0 and similar efforts, and while partnering with other agencies to ensure that these R&D efforts continue after the initial 5 years of CHIPS Act funding expires.

Near-Term Advances: Packaging, Integration, and Other Technologies

Recent months have seen several new initiatives in the semiconductor sector, particularly focusing on packaging and chip integration technologies, which are expected to materialize within 5 years. At the forefront of these efforts are DARPA and NIST. DARPA recently announced the establishment of a Next-Generation Microelectronics Manufacturing (NGMM) center with a budget of $430 million. This center is focused on 3D heterogeneous integration (3DHI). This approach marks a shift from traditional monolithic chip production to a more modular method using chiplets that can be assembled like LEGO blocks. This method offers the potential to disaggregate functions such as memory and processing, thereby significantly boosting performance and lowering cost. DARPA's initiative also includes exploring the integration of photonics and non-silicon electronics into these systems.

Simultaneously, NIST announced the National Advanced Packaging Manufacturing Program (NAPMP), backed by a substantial $3 billion from the CHIPS Act. This program is another major initiative in packaging and integration, covering various elements like materials, tools, thermal management, photonics, and the chiplet ecosystem. The overlapping elements of DARPA and NIST's programs highlight the need for coordination between the two to maximize efficiency and avoid duplicative efforts.

NIST has established the NSTC as a part of the CHIPS Research and Development Office. The NSTC, conceived as a public–private consortium for semiconductor R&D, will involve academia and industry in addressing industry barriers, including workforce development. While funded for a multiyear period, there is uncertainty about the extension of this funding beyond its initial timeline.

DoD has launched the Microelectronics Commons (ME Commons), a 5-year PPP with industry and university hubs, focusing on six key semiconductor technology areas, including broadband technology, artificial intelligence (AI), and quantum technology—funded by the CHIPS Act. The goal of the ME Commons is to bridge the gap between university research and semiconductor production and to enhance U.S. leadership in these fields. Notably, however, this initiative does not address the need for groundbreaking technologies beyond the CMOS era.

Effective coordination among these initiatives is necessary to ensure the seamless exchange of ideas, prevent duplication, and avoid research dead ends, thereby ensuring optimal advancement in the semiconductor field.

Finding 5.1: Four major public–private partnership (PPP) programs have been formed to date by DoD and NIST in semiconductor R&D and in packaging: the Next-Generation Microelectronics Manufacturing program by DARPA, the ME Commons by DoD, NSTC by NIST, and NAPMP by NIST. Given the importance of research advances to retention of a technology leadership role by the United States, it is imperative that the agencies and programs cooperate closely on these programs. Other countries have shown that PPPs work best when they are truly national scale efforts, such as IMEC in Europe or ITRI in Taiwan. The convening authority and ability to work with industry is much simpler when there are fewer major efforts, as opposed to multiple smaller efforts. This has been shown by the international success. Moving forward, there should be fewer efforts and national centers, with perhaps a drive toward only one.

Recommendation 5.1: The Department of Defense (through the director of the CHIPS Coordination Cell) should coordinate with the National Institute of Standards and Technology on its semiconductor-related public–private partnership efforts, including via the Microelectronics Commons, to avoid duplication and ensure the exchange of promising advances.

This coordination might involve merging activities and funding vehicles to avoid duplication and ensure exchange of promising advances. It could also involve personnel exchanges and having laboratory facilities that are open for co-work to reduce barriers to access.

For significant progress to be made in the coming half decade on the initial range of semiconductor technologies, starting with advances on packaging and integration, it will be important to assure that researchers are sharing progress and findings. The four DoD and NIST programs should also assist in lowering another barrier to progress in semiconductor research—the need to provide university semiconductor researchers as well as small and medium-sized companies with access to advanced fabrication facilities.

Finding 5.2: University researchers and start-up companies can play significant roles in supporting semiconductor research advances. However, because semiconductor production facilities, particularly advanced facilities, are so expensive, these institutions lack access to the relevant equipment to undertake research as well as related development, prototyping, and testing. Ongoing and newly created DoD and NIST programs and initiatives offer an opportunity to significantly improve the equipment and technology access problems faced by university researchers and innovative new companies.

There are several ways DoD and NIST can implement collaborations. For example, DoD program managers could be incentivized to use the NSTC facilities that are being stood up as part of the CHIPS Act. They could run programs specifically to use the new facilities that are brought online, potentially even with a reduced rate of operations. To ensure university participation, there could be one or more national conferences that jointly shares results. The CHIPS Act coordination team could review the findings to look for overlaps and promote collaboration. Lastly, there could be budget sharing and interchange of personnel through NATCAST.

Recommendation 5.2: The Department of Defense (through the Office of the Under Secretary of Defense for Research and Engineering and the Office of the Under Secretary of Defense for Acquisition and Sustainment) should actively leverage its own programs and partner with the National Institute of Standards and Technology to connect academic researchers with industry partners and advanced fabrication and packaging facilities.

Post-Complementary Metal-Oxide-Semiconductor Advances

The discussion above highlights DoD's need to focus on long-term advancements in semiconductor technology, particularly beyond the current CMOS technologies. This shift is crucial to achieving significant gains in speed, energy use, and density in semiconductor devices. The International Roadmap for Devices and Systems (IRDS) of the semiconductor sector suggests that the industry's future lies in moving beyond CMOS technologies, and there is a pressing need for systematic efforts to ensure these new technologies are developed and accessible within the next decade, as current scaling methods are expected to reach their limits.

A wide range of CMOS approaches offer possible advancements. Researchers have the flexibility to explore various options to find promising solutions. The Nanoelectronics Research Initiative (NRI) of the Semiconductor Research Corporation (SRC) is engaged in benchmarking these potential approaches against one another to assess viability. Besides hardware innovations, significant efficiency improvements can also be achieved through software performance engineering. This involves restructuring software to enhance computer application speed by removing unnecessary elements and optimizing software for specific hardware features. Additionally, advancements in algorithms that require lower computation levels and hardware streamlining that uses fewer transistors and chip area are other avenues for gains.

While NSTC and the ME Commons may include research in some of these areas, there is need for a more systematic and accelerated research effort in post-CMOS technologies. DoD's role is not just to identify and fund research in these technologies, but also to actively participate in a broader, collaborative research agenda with the industry and other entities like NSTC. The scale of this challenge is beyond what DoD can address alone, underscoring the importance of collaborative efforts in advancing post-CMOS research.

Finding 5.3: As semiconductor fabrication continues to scale ever smaller to achieve speed and energy advances, the physics of CMOS technology is becoming increasingly challenging. In approximately a decade, in order to sustain semiconductor advances to meet the demands of AI, quantum, and more-specialized chips, there will be a need for post-CMOS technologies that advance speed and lower energy consumption through new materials, new architectures, and new computational approaches. DoD will need to take advantage of these post-CMOS R&D initiatives to sustain competitive advantage.

In its quest to assure access to the most advanced manufacturing technology, DoD, via the Office of the Under Secretary of Defense for Research and Engineering (USD R&E), will need to engage in longer-term, early-TRL R&D as well as support device, chip, and package prototyping. This work will be most successful if DoD is a robust partner to NSTC, with sustained investments and clear strategic directions. DARPA's ERI 2.0 could be leveraged strongly to engage industry and university researchers in this work. Similarly, entities like DoD's Office of Strategic Capital, the Defense Innovation Unit, and In-Q-Tel could be leveraged strongly for prototyping and scale-up support.

Recommendation 5.3: The Department of Defense (through the Office of the Under Secretary of Defense for Research and Engineering) should focus on long-term scientific research in disruptive semiconductor technologies, including post-CMOS technologies, and ensure broader access to prototyping facilities for academic researchers and small to medium-sized firms.

ARTIFICIAL INTELLIGENCE ADVANCES

It is difficult to overstate the potential value of emerging AI and machine learning (ML) technologies for military use. As noted throughout the report, AI semiconductors are critical for maintaining technological superiority in defense capabilities, enabling advanced systems for national security, cyber defense, and autonomous military operations. Ensuring the security and innovation of these semiconductors is essential for safeguarding sensitive information and enhancing the strategic advantage of the U.S. military. It is very likely that the nations that dominate the AI landscape will also have global national security leadership.

Finding 5.4: In the global competition surrounding the use of AI, the entity that controls its AI hardware and software supply chain will have a significant advantage in future conflicts. While DoD anticipated the arrival of the AI revolution, the necessary investments to create the hardware and software required were made by the private sector. Today, DoD requires access to AI technologies while ensuring reliable and resilient supply chains.

Recommendation 5.4: The Department of Defense (DoD) (through the Chief Digital and Artificial Intelligence Office) should swiftly incorporate U.S.-origin commercially developed artificial intelligence (AI) models, chips, and infrastructure, and develop custom AI technologies for defense purposes, to maintain leadership in this transformative technology. DoD should also track leading-edge AI technologies and align DoD efforts with commercial interests to mitigate supply chain risks.

ADDRESSING DESIGN BARRIERS

The challenge of high costs of designing application-specific integrated circuits (ASICs), which are crucial for achieving technological superiority in military systems, has only grown more acute over time, and it requires the development of new design tools to enable the creation of affordable, high-performance, low-volume custom ASICs (see Chapter 2). Although the commercial electronic design automation industry is developing such tools, the rate of investment is insufficient to meet DoD's specific needs. Advancements in software and AI technology offer the potential for rapid progress in this area.

The CHIPS Act, focused on production facilities, did not address the critical issue of design costs, and reducing these costs remains a vital barrier to exploiting technological opportunities afforded by access to advanced chip manufacturing technology. This challenge is not unique to DoD, but it is also faced by start-ups, small and mid-sized firms, and large companies seeking to develop applications for moderate-sized markets using advanced chips. Therefore, a collaborative R&D initiative led by DoD, working with industry to support new lower-cost design approaches, would have significant benefits for both the defense sector and the wider industry.

Human resource constraints are another factor increasing costs and limiting the integration of advanced semiconductors into new DoD systems. The application of AI and ML software tools may reduce the need for large teams of software engineers to do chip design and development, and it may also reduce the (currently steep) barriers to entry into this discipline, thus addressing these constraints.[1] This approach is further supported by recent DARPA semiconductor research and prototyping initiatives that include a focus on design automation and simulation software, priming the conditions for a larger opportunity that translates these concepts into practice.

---

[1] E. Sperling, 2024, "Preparing for an AI-Driven Future in Chips," Semiconductor Engineering, February 1, https://semiengineering.com/preparing-for-an-ai-driven-future-in-chips.

Finding 5.5: With DoD requiring custom chips for its specialized needs, with the cost of chip design accelerating, and with the need for specialized chips to meet different tasks (such as for AI and quantum) growing, DoD is facing design costs that could affect its security missions. Major segments of industry also have an interest in lowering design costs. Development of AI and other tools may in turn enable a new suite of lower-cost design options and approaches. Therefore, a flagship program to significantly reduce design costs appears in order. If the United States does not undertake this, it could cede its design leadership to others ready to undertake it. An investment in development and deployment of more advanced chip design software tools is a natural and logical complement to DoD's current ME Commons initiative.

Recommendation 5.5: The Department of Defense (through the Office of the Under Secretary of Defense for Research and Engineering) should organize a significant initiative, in partnership with industry, to leverage artificial intelligence and machine learning tools to substantially reduce the time and cost of application-specific integrated circuit design and software development for defense needs.

Recommendation 5.6: The Department of Defense (through the Office of the Under Secretary of Defense for Research and Engineering) should build skills internally for modern chip design, partner with commercial-sector experts designing chips at advanced nodes, and tap into the Department of Commerce–supported supply chains for manufacturing necessary custom chips.

Recommendation 5.7: The Department of Defense (DoD) (through the Office of the Under Secretary of Defense for Research and Engineering) should create a flagship unit with expertise in hardware cybersecurity and modern chip design that DoD program managers can access as needed. DoD should explore embedding design teams from this unit within companies developing leading-edge chip designs to increase DoD's technical expertise.

ADDRESSING BUREAUCRATIC AND REGULATORY BARRIERS

DoD is facing significant challenges in adopting advanced chip technologies, primarily owing to internal security requirements and the reliance on trusted foundries, as well as regulations like ITAR (International Traffic in Arms Regulations) and EAR (Export Administration Regulations). Complying with these requirements requires a sophisticated understanding of complicated regulations that are administered by multiple agencies and that change frequently. Export controls are administered by DOC and its Bureau of Industry and Security (BIS), which regulate the export, reexport, and in-country transfer of items on the Commerce Control List (most commercial and some military goods). This includes dual-use items that have civil applications as well as military applications. Before exporting an item, a company must determine whether a license is required based on the nature of the item, its destination (both geographic location and identity of the recipient), and its intended use. ITAR, on the other hand, is administered by the Department of State's Directorate of Defense Trade Controls (DDTC) and regulates the export of defense items, including data relating to defense items, and defense services. The U.S. Munitions List (22 CFR 121.1) sets out those items subject to ITAR, but it includes numerous ambiguities and catch-alls, often making it difficult to interpret. A registered company must obtain a license or written agreement from DDTC before exporting any ITAR-controlled items or services. Businesses are expected to determine how items should be classified, but failure to comply with ITAR and EAR carries severe penalties, including fines, imprisonment, and debarment. If an exporter has doubt about the proper classification of an item, it must submit a written request called a Commodity Jurisdiction, and DDTC will provide advice on whether something is a dual-use item subject to EAR or a defense item regulated by ITAR. Further complicating the situation, in October 2022 the Commerce Department amended EAR by announcing new regulations targeting China's ability to acquire semiconductor manufacturing technologies and advanced computer chips. Those controls were expanded in December 2023, further limiting exports of advanced computer chips and associated manufacturing technologies to China and to Chinese entities located in other countries.

This system has led to inefficiencies in acquiring custom advanced chips, impacting national security readiness. The recent development of a new jam-resistant GPS capability by the Air Force, a billion-dollar project, illustrates the regulatory hurdles in using advanced chips. The project design was completed in a short time, but navigating through regulatory barriers to justify the use of U.S. facilities took years.

The Microelectronics Quantifiable Assurance (MQA) system discussed in Chapter 3 aims to assure the confidentiality, integrity, and availability of microelectronics while accessing the commercial supply chain. However, transitioning from the trusted foundry model to an evidence-based assurance approach has been behind schedule owing to various challenges, including difficulties in developing and staffing new processes and the impact of coronavirus disease 2019 (SARS-CoV-2). This delay has put DoD 3 years behind the requirements set by the fiscal year (FY) 2020 NDAA to establish trusted supply chain and operational security standards. DoD faces a critical trade-off: to secure access to the latest microelectronics, it must alter its approach to managing a "secure" supply chain and align more closely with commercial industry practices. DoD should do whatever it can to align with domestic manufacturing of components, and leverage DOC's investment to increase resilience in supply chains.

Finding 5.6: DoD requires access to the most advanced chips to keep up with technology advances it must have to meet national security needs. To access cutting-edge chips for its customized needs, it will need to rely on commercial firms, and develop an evidence-based chip quality and security assurance system for a trusted supplier ecosystem. There are significant barriers to accessing the newest U.S.-based plants that will be funded by DOC, based on current DoD rules. This mismatch between security concerns, DoD's semiconductor needs, and economic resilience is a weakness within the country and stops the United States from having world-class national security efforts.

Recommendation 5.8: The Department of Defense (DoD) (through the Office of the Under Secretary of Defense for Acquisition and Sustainment) should accelerate efforts to implement an evidence-based assurance system to ensure fast and secure access by DoD programs to advanced commercial semiconductors, and also simplify bureaucratic processes and update relevant DoD instructions and policies in support of this effort.

This should be a major, high-priority effort, with appropriate resources allocated to develop and enhance a Microelectronics Evidence-Based Assurance System for secure semiconductors. This system should guarantee DoD's ability to rapidly acquire advanced, commercially sourced, customized chips for its systems. Once established, program managers for technology projects should have the authority to obtain chips without further review, expediting the integration of advanced chips into systems. DoD should fully implement the requirements of Section 224 of the National Defense Authorization Act for Fiscal Year 2020, while ensuring alignment with commercial practices to avoid creating unnecessary gaps with manufacturers. This includes (1) collaborating with commercial process design kits wherever possible, instead of establishing a separate set of standards; (2) updating DoD instruction 5200.44; (3) publishing a new policy to implement the evidence-based assurance method; (4) updating the Joint Federated Assurance Center Charter and Concept of Operations to develop a process for prioritizing the evidence-based assurance efforts of supporting DoD laboratories; and (5) identifying the resources required to support the National Security Agency's role in threat analysis for evidence-based assurance, or identifying another DoD organization capable of performing this role. While these have been studied or implemented before, the current focus should be on simplifying, minimizing bureaucratic hurdles, and maximizing access to the supply chain created by the CHIPS Act funding. This alignment of efforts between DOC and DoD is crucial.

Recommendation 5.9: The Department of Defense (DoD) (through the Office of the Under Secretary of Defense for Acquisition and Sustainment) should review and update policies that limit it from manufacturing in commercial facilities. DoD should also increase the use of domestically manufactured chips where possible and simplify procurement processes to streamline access to semiconductor suppliers.

Finding 5.7: Current ITAR and EAR regulate the export of defense-related technologies, information, and services, including information conveyed to (and therefore work performed by) nonpermanent residents. Given the international nature of the semiconductor sector professional workforce, these regulations, as currently interpreted, are limiting DoD's access to an important talent pool. Accordingly, a new strategy is needed. In addition, there are signals that application of environmental reviews, such as through the National Environmental Policy Act may cause significant construction delays in new fabrication plants (fabs). Further legislative authority may be required to allow DoD to better "onshore" and "friend-shore" required manufacturing activity.

Recommendation 5.10: The Department of Defense (DoD) (through the Defense Technology Security Administration) should collaborate with relevant organizations to reduce the administrative burdens and improve the timeliness of decisions related to International Traffic in Arms Regulations, Export Administration Regulations, and the National Environmental Policy Act. DoD should consider forming a task force to review and reform these regulations.

ADDRESSING BARRIERS TO MANAGING INTELLECTUAL PROPERTY RIGHTS

Effective management of intellectual property (IP) rights presents another challenge that DoD must resolve in a PPP, including not just patent rights, but also data rights in computer software and technical data, mask works, and trade secrets and related know-how. Consideration must be given to ownership of IP created during the PPP, government rights to IP created with the direct or indirect use of government funding, requirements for sharing or transferring IP to other PPP participants and/or to the public, and restrictions on publication and other dissemination of information. Four general models can be utilized by DoD, depending on the technology readiness level (TRL) and intended use: (1) open access to IP; (2) sharing IP between members of the PPP only; (3) retention of IP rights by each PPP to IP created by the participant; and (4) when necessary, federal government ownership of IP created during the PPP. Each is discussed in turn below.

Open Access to Intellectual Property

PPPs are sometimes characterized as a basic form of institutionalized knowledge sharing, directed to creating a "commons" where information is collectively owned and managed. Other commentators have argued that PPPs should recognize a form of patent fair use in order to promote collaborative research.[2] However, in the context of a PPP intended to strengthen DoD supply chain for semiconductor devices, open access to IP has the potential to dramatically undermine the very purpose of the PPP and could potentially lead to sharing of information with adversaries. Instead, DoD must create a model that preserves the rights of private partners and their ability to commercialize inventions created in the PPP.

Open access to the general knowledge and basic skills of industry workers (the level of ordinary skill in the art) requires a different analysis, because such knowledge and skills are likely already in the public domain. One area where open IP has been touted is in the context of workforce training programs. Intel has created Semiconductor Education and Research Programs with a group of Ohio universities and community colleges, which it describes as an "open IP" model.[3] In some instances, Intel Laboratories conducts its own research first, then explores ways in which that research can be augmented by graduate students working at several universities (known as the Academic Mindshare program). Another effort will provide experiential opportunities for students, including a 1-year certificate program for technicians that makes them eligible for entry-level positions at Intel or another semiconductor company. In these settings where little or no IP will be created and the focus is on training technicians and other lower-level employees, the concept of open IP takes on a different meaning.[4]

Consortium Models for Sharing Information

The Semiconductor Manufacturing Technology consortium (SEMATECH) is one example of a consortium where IP was shared between its members and DoD (see Box 5-2). It was considered to be an extremely successful consortium for developing advanced manufacturing technology, due at least in part to the sharing by member companies of precompetitive information about their manufacturing

---

[2] See, for example, L.S. Vertinskey, 2015, "Patents, Partnerships, and the Pre-Competitive Collaboration Myth in Pharmaceutical Innovation," *UC Davis Law Review* 48:1509 (writing in the context of pharmaceutical PPPs), and other references cited therein.

[3] S. Venkataramani, Intel Labs, presentation on November 28, 2023.

[4] Ms. Venkataramani noted that sponsored research agreements are available if needed.

BOX 5-2
SEMATECH

SEMATECH, Inc., was formed in August 1987 as a nonprofit corporation to conduct research and development intended to provide the U.S. semiconductor industry with domestic capabilities to position it as a world leader in semiconductor manufacturing. It was a consortium of 14 major U.S. semiconductor and computer companies, where each company assigned technical employees ("assignees") to participate in SEMATECH's research and development (R&D) programs.[a] The memorandum of understanding (MOU) between SEMATECH and the Department of Defense (DoD) contained two separate terms addressing ownership and use of intellectual property.

First, the MOU required SEMATECH to take all steps necessary to maximize the expeditious and timely transfer of technology developed and owned by SEMATECH to its participants, in accordance with the agreement between SEMATECH and its participants.[b] As a result, inventions created by the assignees were patented by SEMATECH, and then transferred to participants.[c] Second, even though DoD was not a formal member, apparently because of its financial contributions, the MOU provided that DoD was permitted to use intellectual property, trade secrets, and technical data owned and developed by SEMATECH in the same manner as a participant in the consortium. DoD was also permitted to transfer such intellectual property, trade secrets, and technical data to DoD contractors for use in connection with DoD requirements, provided that DoD did not transfer IP to any person for commercial use.[d]

---

[a] Each company signed a participation agreement requiring the company to support SEMATECH at certain funding levels for an initial period of 4 years. A member company's annual financial contributions were calculated based on its prior year's sales of semiconductor devices (for those members who produced electronic equipment, the value of its semiconductor purchases in the previous year was used instead). A company's financial contributions then determined the number of technical employees it could assign to SEMATECH, along with the extent of technology and associated know-how that could be transferred to the company. See U.S. General Accounting Office, 1989, "The SEMATECH Consortium's Start-up Activities," November, p. 11.

[b] SEMATECH, 1988, "Memorandum of Understanding," May 12, p. 4.

[c] P.N. Dunn, 1997, "SEMATECH's Burgeoning Patents May Be Future Revenue Generator," *Solid State Technology* 40(1).

[d] SEMATECH, 1988, "Memorandum of Understanding," May 12, pp. 3-4. See also 15 U.S.C. § 4602(b).

processes and equipment.[5] It is unclear, however, whether this model would attract participation today for anything beyond basic research and pre-commercialization efforts (TRL 1–3). Companies appear less interested in working collaboratively because they want to preserve their IP rights to both background IP and new developments, and they frequently prioritize profits over the collective good of the

---

[5] U.S. General Accounting Office, 1992, "Lessons Learned from SEMATECH," September, p. 4. The GAO report further discussed the importance of SEMATECH as a forum for communications between members. Specifically, through SEMATECH discussions, industry members found that they were protecting similar trade secrets and attempting to address similar manufacturing problems. *Id.* at p. 8.

industry or the needs of DoD.[6] Start-ups and small firms, on the other hand, may show more willingness to collaborate and share developments, particularly if they can license background IP that is necessary to their R&D efforts.

Retention of Intellectual Property Rights by Public–Private Partnership Participants

In order for any PPP to be successful, it must allow for private ownership of any IP developed during the partnership that relates to advanced TRLs. In a typical situation where one private party contracts with a second private party for research services, the paying party expects to own any IP created as a result of the relationship. In dealing with the government, however, the relationship is inverted: while the government provides funding, the private party conducting the research expects to own any resulting IP. Daniel Armbrust, former president and chief executive officer of SEMATECH, acknowledged that ownership of IP is critical, particularly for start-up companies. Indeed, he described it as their "lifeblood."[7] Private ownership of IP is recognized and endorsed by the Bayh-Dole Act and federal regulations governing data rights.

The federal Bayh-Dole Act[8] applies to inventions arising from federally supported R&D and specifically authorizes private parties to own patents resulting from such work. The act expresses the policy and objective of Congress to, inter alia, promote the commercialization of and public availability of inventions made in the United States by U.S. industry and labor, as well as to ensure that the government obtains sufficient rights in federally supported inventions to meet the needs of the government and protect the public against nonuse or unreasonable use of such inventions.[9] The act originally applied only to small businesses, universities, and other nonprofit organizations; however, in 1983, President Reagan

---

[6] That concern is reinforced by the experience of IMEC, a nanoelectronics center located in Leuven, Belgium, that has been involved in collaborative R&D for more than 39 years. IMEC serves as a bridge between universities, government, and industry, although it receives 75 percent of its funding from industry sources. Much of IMEC's work is at the precompetitive stage, where both costs and IP are shared and research results are pooled. An IMEC member company pays a background fee to obtain IP created prior to joining, followed by annual fees for access to future developments. See L. Lauwers, IMEC, presentation to the committee, July 18, 2023. Mr. Lauwers explained that IMEC also provides research services at more competitive stages, where IP can be owned exclusively by the company paying the costs and fees associated with the work. This is separate from the consortium program.

[7] See D. Armbrust, Silicon Catalyst, presentation to the committee, September 26, 2023. Mr. Armbrust suggested that having an IP template for a PPP is unlikely to be successful, however. He observed that every situation is different: each one has novel issues specific to the stakeholder, funding, and technologies involved. Standard IP terms are therefore not likely to be useful, and flexibility is required.

[8] 35 U.S.C. § 200 *et seq.*, enacted 1980.

[9] 35 U.S.C. § 200.

extended its benefits to all contractors, including large businesses and for-profit organizations.[10]

The Bayh-Dole Act provides that a party to a funding agreement may, within a reasonable time after disclosure of a subject invention, elect to retain title to that invention.[11] In order to invoke the process, the contractor must disclose a subject invention to the federal agency within a reasonable time after it becomes known to contractor personnel responsible for the administration of patent matters. The contractor must then make a written election to retain title within 2 years after disclosure to the federal agency, except in those cases where the inventor's 1-year filing window under Section 102(b) would end before the end of the 2-year election period. The contractor must then file a patent application prior to the expiration of the 1-year period allowed by Section 102(b) and then file corresponding patent applications in other countries in which it wishes to retain title. The federal government may receive title to any subject inventions in the United States or in other countries in which the contractor has not filed patent applications on the subject invention within such times.[12]

However, even when a contractor elects to retain title to a subject invention, the federal government is not devoid of rights in the invention it funded. Instead, the federal agency receives a nonexclusive, nontransferable, irrevocable, paid-up license to practice or have practiced for or on behalf of the United States any subject invention throughout the world.[13] For example, in an instance where DoD funds the development of a new ASIC, its license rights allow DoD to share the subject IP with another contractor for purposes of having the second contractor make the same ASIC for DoD. DoD should not be required to pay twice for the creation of the same invention.[14] While DoD is sometimes forced to pay twice for certain chips

---

[10] See Presidential Memorandum dated February 18, 1983. An exception can be made where (1) it is necessary to obtain an agreement with a uniquely qualified contractor or (2) the award involves co-sponsored, cost sharing, or joint venture research and development and the performer, co-sponsor, or joint venturer is making substantial contribution of funds, facilities, or equipment to the work performed under the award.

[11] 35 U.S.C. § 202(a). The term "subject invention" means any invention of a contractor conceived or first actually reduced to practice in the performance of work under a funding agreement. The term "funding agreement" means any contract, grant, or cooperative agreement entered into between any federal agency and any contractor for the performance of experimental, developmental, or research work funded in whole or in part by the federal government. See 35 U.S.C. § 201 (definitions).

[12] 35 U.S.C. § 202(c).

[13] 35 U.S.C. § 202(c)(4). The government also has the ability to order that an invention be kept secret and can withhold publication of an application or the grant of a patent, when the government believes that disclosure could be detrimental to national security. See 35 U.S.C. § 181.

[14] Any U.S. patent that issues must include a statement indicating that the invention was made with government support and that the federal government has certain rights in the invention. See 35 U.S.C. § 202(c)(6).

or capabilities, that duplication of costs likely does not arise as a result of the first contractor's election to retain title to a federally funded invention. More often, it results from pressure by subsequent contractors to be paid the maximum amount possible to fund creation of a DoD system or device,[15] or from DoD's poor record-keeping about those inventions in which it has rights.[16] DoD apparently does a very poor job of tracking inventions resulting from federal funding, in which it now has nonexclusive rights. It is recommended that DoD conduct a comprehensive audit of patent and other IP rights in which it may presently have an interest. In the future, DoD should implement a central recordkeeping system to track invention disclosures, patent applications submitted to the U.S. Patent and Trademark Office and to foreign patent offices, and issued patents resulting from federal funding.

To the extent practicable, federally funded IP should be manufactured in the United States. The Bayh-Dole Act expresses a strong preference for U.S. industry. A business that receives title to a federally funded invention is precluded from giving an exclusive right to use or sell the invention in the United States unless the products will be manufactured substantially in the United States, unless domestic manufacture is not commercially feasible.[17] Similar requirements apply when the government retains ownership of an invention that it funds. The government should also place appropriate restrictions on the sale of federally funded IP to foreign businesses and foreign investors.

The federal government also retains "march-in" rights in federally funded inventions, allowing it to grant an exclusive or nonexclusive license to a third party when a contractor fails to take steps to achieve practical application of the invention within a reasonable time or where such action is necessary to alleviate health or safety needs that are not being satisfied by the contractor or its licensees.[18] To date, the federal government has never exercised its march-in rights.[19]

---

[15] But see FAR § 27.306 and DFARS § 252.227-7038(i), regarding rights of the U.S. government in background inventions. The government may claim that its license to use a subject invention has little value, unless it also acquires a license to the preexisting technology.

[16] DoD apparently does a very poor job of tracking inventions resulting from federal funding, in which it now has nonexclusive rights. It is recommended that DoD conduct a comprehensive audit of patents and other IP rights in which it may presently have an interest. In future, DoD should implement a central recordkeeping system to track invention disclosures, patent applications submitted to the USPTO and to foreign patent offices, and issue patents resulting from federal funding.

[17] See 35 U.S.C. § 204.

[18] 35 U.S.C. § 203(a) (December 12, 1980; last amended January 4, 2011).

[19] Comment of U.S. Federal Trade Commission to the Department of Commerce, National Institute of Standards and Technology, "Draft Interagency Guidance Framework for Considering the Exercise of March-in Rights," at 2 (February 6, 2024), available at Comment of the United States Federal Trade Commission (ftc.gov). See also Congressional Research Service, *March-In Rights Under the Bayh-Dole Act* at 8 (August 22, 2016).

Recently, however, the Biden administration proposed using march-in rights to lower drug costs, and NIST issued a request for comments on use of march-in rights and forced licensing more generally.[20]

Ownership and use of data rights must also be addressed, because they apply to both trade secret information and works subject to copyright. The federal government obtains rights in technical data, including a copyright license, under an irrevocable license granted (or obtained for) the government by the contractor. The contractor retains all rights in the data not granted to the government. The scope of the license is generally determined by the source of funds used to develop the data.[21] To encourage contractors to offer or use commercial products to satisfy military requirements, contractors shall generally not be required to (1) furnish technical information related to commercial items or processes that is not customarily provided to the public or (2) relinquish to, or otherwise provide, the government rights to use, modify, reproduce, release, perform, display, or disclose technical data pertaining to commercial items or processes except for a transfer of rights mutually agreed upon.[22]

Accordingly, for commercial items, DoD acquires only the technical data customarily provided to the public with a commercial item or process, with three exceptions. Those exceptions include technical data that (1) are form, fit, or function data; (2) are required for repair or maintenance of commercial items or processes, or for the proper installation, operating, or handling of a commercial item; or (3) describe modifications made at government expense to a commercial item or process in order to meet the requirements of a government solicitation.[23]

For noncommercial items, DoD acquires only the technical data and rights in that data sufficient to satisfy DoD's needs.[24] The standard license rights that a licensor grants to the government are unlimited rights, government purpose rights, or limited rights.[25] Generally speaking, DoD acquires unlimited rights in technical data only when an item, component, or process has been developed exclusively with government funds. DoD acquires only lesser, government purpose rights in technical data relating to an item, component, or process developed with mixed

---

[20] 88 Fed. Reg. 85593 (December 8, 2023). The Federal Trade Commission issued a comment supporting "an expansive and flexible approach" to march-in rights, including use of march-in rights to lower drug prices.

[21] DFARS § 227.7103-4(a).

[22] DFARS § 227.7102-1(b).

[23] DFARS § 227.7102-1(a).

[24] DFARS § 227.7103-1.

[25] DFARS § 227.7103-5.

funding. For technical data relating to an item, component, or process developed exclusively with private funds, DoD receives only limited purpose rights.[26]

Government purpose rights are rarely used and present an artificial barrier to working with nontraditional entities. DoD should consider removing government purpose rights for semiconductor engagements and replacing them with a U.S. manufacturing preference. Such a preference is already incorporated in the Bayh-Dole Act, which provides that no small business or nonprofit organization that receives title to a subject invention shall grant any person the exclusive right to use or sell the invention in the United States unless that person agrees that any products embodying the invention will be manufactured substantially in the United States.[27]

By allowing private parties to retain ownership of IP created during a PPP (including patents, copyrights, and trade secrets), both large and small companies will be incentivized to participate in the partnership because they will emerge with intellectual assets that build value in their companies. Private parties must be able to commercialize the technologies they create in the context of a PPP. The costs of participating in these efforts are so high (both the direct cost sharing and associated operating costs) that private partners must have the ability to use IP created out of a PPP.[28] That need is further reinforced by NIST's request for comments regarding the NSTC, where respondents clearly stated they need to know how IP rights are going to be allocated before they can make any decision about whether to participate in a PPP.[29] Remember that corporations have a duty to investors and shareholders to maximize profits.

Government Ownership of Intellectual Property

Government ownership of IP should be restricted to those extreme situations where ASICs are being developed solely for use by DoD and the chips have no commercial application. Because the objective of this report is to consider PPPs that benefit both DoD and dual-use needs, government ownership should arise

---

[26] DFARS § 227.7103-5. In unusual situations, these standard rights may not satisfy the government's needs or the government may be willing to accept lesser rights in data in return for other considerations. In those cases, a special license may be negotiated.

[27] See 35 U.S.C. § 204. Note though that in individual cases, the requirement can be waived by the federal agency that funded the invention if the patent owner shows that reasonable but unsuccessful efforts were made to grant licenses to persons that would be likely to manufacture substantially in the United States or that under the circumstances domestic manufacture is not commercially feasible.

[28] Presentation by Mukesh Kare, IBM, September 5, 2023.

[29] NIST, 2022, "Incentives, Infrastructure, and Research and Development Needs to Support a Strong Domestic Semiconductor Industry: Summary of Responses to Request for Information," Special Publication NIST SP 1282, https://doi.org/10.6028/NIST.SP.1282.

only rarely. In those unusual circumstances, DoD would own some or all of the IP resulting from a PPP, and it could then transfer that technology to government contractors on a limited basis, subject to appropriate security clearances and controls. Government ownership is authorized by the Bayh-Dole Act, which recognizes that in some circumstances a contractor may not have a right to retain title in a subject invention, and the federal government will own inventions made under a federal funding agreement. Government ownership is required where the contractor is not located in the United States, does not have a place of business located in the United States, or is subject to the control of a foreign government. In exceptional circumstances, a funding agreement may also restrict or eliminate the right to retain title when the federal agency determines that such restriction or elimination of rights will better promote the policy and objectives of the act. Similarly, if a federal agency is authorized to conduct foreign intelligence or counterintelligence activities, the right to retain title may be restricted or eliminated as necessary to protect the security of those activities.[30] The data rights provisions in the Defense Federal Acquisition Regulation Supplement (DFARS) likewise authorize DoD to acquire unlimited rights in technical data relating to an item, component, or process that has been developed exclusively with government funds.[31]

Finding 5.8: Consortium models that require broad sharing of background IP and IP developed during a PPP are unlikely to be attractive to industry members beyond TRL 1–3. At more advanced TRLs, industry members will expect to own IP created during the PPP in order to preserve the right to commercialize inventions and maximize profits. Private ownership of IP is expressly authorized by the Bayh-Dole Act and the federal regulations governing data rights.

Recommendation 5.11: The Department of Defense (DoD) (through the Office of the Under Secretary of Defense for Research and Engineering) should manage intellectual property (IP) rights based on technology readiness levels and intended end use, recognize private ownership of IP in the majority of cases, reserve government ownership only for extreme situations, conduct an audit to track DoD's existing IP rights, and create a central, searchable records system.

---

[30] 35 U.S.C. § 202(a). A final restriction relates to the operation of government-owned, contractor-operated facilities of the Department of Energy primarily dedicated to DOE's naval nuclear propulsion or weapons related programs.

[31] *Id.* In unusual situations, these standard rights may not satisfy the government's needs or the government may be willing to accept lesser rights in data in return for other considerations. In those cases, a special license may be negotiated.

ADDRESSING MANUFACTURING BARRIERS

Chapter 3 summarizes the significant challenges in manufacturing processes, especially as production costs have risen sharply owing to the demands of scaling in semiconductor manufacturing. This issue is not unique to DoD but is common across industries and is linked to market size constraints. DoD's need for ongoing access to leading-edge chips necessitates encouraging advanced manufacturing capabilities within the U.S. industry. While the majority of semiconductor production will remain in Taiwan, South Korea, and other countries, having some capacity in the United States is crucial for DoD's future technology needs. DoD's Office of Strategic Capital could become a significant player in the semiconductor ecosystem if it increases its financing level.

With the CHIPS Act being a 5-year authorization and no guarantee of extension, DoD, in collaboration with DOC, needs to consider extending financing mechanisms for advanced semiconductor facilities beyond the act's timeframe. The issues of U.S. chip leadership and the need for government financing to ensure production capability in the United States are expected to persist beyond this period. Therefore, proactive planning for continued manufacturing support is essential for DoD to maintain access to the advanced semiconductor technologies critical for national defense.

Finding 5.9: U.S. firms have lost leadership in producing the most advanced chips, yet DoD's future needs for advanced technologies dictate that it has access to at least one cutting-edge semiconductor fab in the United States, as well as other domestically located key elements of the semiconductor ecosystem. Since the CHIPS Act is only a 5-year authorization for support for chip production and DoD's needs will continue long beyond that term, it is important that financing support tools (e.g., subsidies, loan guarantees, and tax credits) be extended to support advanced semiconductor facility construction in the United States on a continuing basis.

Recommendation 5.12: The Department of Defense (DoD) (through the director of the CHIPS Coordination Cell) should act to secure longer-term federal funding support for the semiconductor sector, beyond the term of the CHIPS and Science Act of 2022, to ensure DoD has continual access to the most advanced semiconductor technologies in the world. DoD should also ensure that these technologies can be sourced from diverse foundries located in the United States and friendly nations.

DoD needs continuous access to advanced semiconductor foundries capable of both high- and low-volume manufacturing to meet its specific requirements, including in emergency situations. However, these foundries, to be commercially competitive, cannot primarily cater to DoD needs alone. This situation necessitates that DoD adapt its security controls in hardware supply chains and accept some level of risk that comes with commercial partnerships. DoD's approach should focus on foundries that serve both defense and commercial markets (dual use) that are capable of producing both high and low volumes and maintain a competitive environment with multiple producers in the United States and allied countries. This approach acknowledges the realities of global supply chains and the necessity of international collaboration.

DoD has initiated programs like Rapid Assured Microelectronics Prototypes (RAMP) and RAMP-C to develop secure design and prototyping capabilities without relying on closed security architectures. These programs have enabled partnerships with leading firms like Intel, Boeing, and Northrop Grumman. In addition, DoD should consider PPP models to ensure access to advanced semiconductor manufacturing capacity. These models could include government loan guarantees combined with direct contributions, similar to those used in the private semiconductor industry and outlined in the CHIPS Act.

DoD also needs to prepare for future technology evolution, following the current period of support by the CHIPS Act. This preparation will necessarily involve both R&D and capital investments for leading-edge facilities and ensuring access to these facilities. The federal government's role in financing advanced fabrication facilities is crucial, as seen from the past two decades of global semiconductor manufacturing.

Finding 5.10: Because DoD has a critical future need for advanced chips, if federal government financing is continued past the 5-year authorization of the CHIPS Act, it will be important that DoD leverage collaborative agreements to assure access for its advanced chip needs as part of such financing.

Recommendation 5.13: The Department of Defense (DoD) (through the director of the CHIPS Coordination Cell) should act to ensure that federal capital and financing supports for the semiconductor sector are tied to agreements that guarantee it access to advanced semiconductors. The DoD Office of Strategic Capital should also prioritize financing for critical technology scale-up activities.

Finding 5.11: DoD has a specific challenge with speed of execution for both design and manufacturing. The Air Force study of the quantifiable assurance development effort[32] highlights the substantial gap in clarity that can slow a

---

[32] V. Coleman, 2023, "NDAA 2023 Mandated Independent Review of USD (R&E) Microelectronics Quantifiable Assurance Effort," Department of the Air Force, August 3, Version 1.0, https://www.af.mil/Portals/1/documents/2023SAF/MQA_Report.pdf.

program by years. The lack of manufacturing options, and clarity in what is allowed in existing manufacturing options, has added to the burden of making complex chip designs. Layers of decision-making around manufacturing (trusted foundry, ITAR, EAR) have slowed the process of chip design and manufacturing. In lieu of the ability to make its own custom ASICs in a timely manner, DoD often utilizes commercial chips (often created overseas).[33]

Recommendation 5.14: The Department of Defense (DoD) (through the Office of the Under Secretary of Defense for Acquisition and Sustainment) should lower barriers to utilizing domestic manufacturing entities, minimizing complex bureaucracy, regulations, and requirements, to obtain the custom chips that DoD needs. The process for determining where DoD's manufacturing can be performed should have short response times so as to avoid unduly hindering DoD's technology programs.

Finding 5.12: U.S. industry is far ahead of the U.S. government in semiconductor manufacturing. It would be self-defeating for DoD to create a siloed infrastructure for DoD use alone.

Recommendation 5.15: The Department of Defense (DoD) (especially the Chief Digital and Artificial Intelligence Office) should prioritize partnering with industry over creating custom solutions, utilizing commercial technology wherever possible. DoD should partner closely with U.S. companies to creatively and nimbly adopt emerging technologies for defense purposes, with immediate urgent attention to artificial intelligence and potential superintelligent systems.

Only by collaborating with the commercial entities that are creating the most advanced, next-generation, large-scale AI systems will DoD develop the needed understanding to use them as needed, and before adversaries do so. DoD will benefit from partnering with industry leaders to ensure that the best-in-class AI tools are resilient, secure, and composed of U.S.-centric software and hardware, most especially as we head toward creating superintelligent systems.

---

[33] The preceding paragraph was modified following the release of the report to clarify the specific chips, ASICs, to which it refers.

ADDRESSING THE BREADTH OF THE DEPARTMENT OF DEFENSE'S SEMICONDUCTOR CHALLENGES

As discussed in the preceding chapters, DoD faces significant challenges to develop new microelectronics-based systems, maintain existing electronic technologies across its sprawling enterprise, and remain at the forefront of new semiconductor technologies. Newer organizations such as the Defense Innovation Unit (DIU) have attempted to expedite the adoption of technologies like AI and machine learning by leveraging commercial technologies.

Finding 5.13: DoD's major defense platforms can last for decades, and the microelectronics require regular updating to remain current and effective. In parallel, DoD needs to be at the forefront of new technology developments in areas such as AI, quantum information science, and advanced data management to remain competitive with other militaries. Clear competitive pressures also accelerate the timelines for DoD to adapt its existing platforms to the latest electronics advances.

Finding 5.14: Also, as DoD works to modernize its various systems, special care is needed to avoid inclusion of fairly mundane legacy semiconductor chips for which the manufacturing may have shifted from allied nations to China during recent years, which could lead to security vulnerabilities. DoD lacks a comprehensive semiconductor strategy to accelerate utilization of the very latest technology in its existing and new systems and platforms.

Recommendation 5.16: The Department of Defense (through the Office of the Under Secretary of Defense for Acquisition and Sustainment, in partnership with the service branches) should develop an overarching microelectronics strategy for research, development, procurement, sustainment, and modernization.

6

General Principles for the Department of Defense (DoD) to Follow for Accelerating the Adoption of Disruptive Technologies to Benefit DoD and Commercial Needs

Given previously described barriers and challenges and corresponding specific recommendations, this chapter will provide general and enduring principles for the Department of Defense (DoD) to drive change. This chapter supplements the report's findings and recommendations with general principles to guide DoD efforts beyond the specific recommendations in Chapters 4 and 5.

PRINCIPLE 1: BE A FAST FOLLOWER

DoD should strive to be a fast follower in rapidly adopting and incorporating into systems new microelectronic technologies developed by commercial industry. Here the committee is considering mainstream logic, memory, and analog integrated circuit process technologies that are advanced by industry leaders such as Intel, GlobalFoundries, Micron, Texas Instruments, Analog Devices, Nvidia, and others. DoD cannot hope to match the capabilities of those companies but should instead strive to adopt those technologies for DoD use very shortly after they are available for commercial purposes. In some cases, DoD may need to work with those companies to develop derivative versions of their technologies to better match DoD needs. Even for the development of emerging mainstream technologies such as artificial intelligence (AI) and quantum computing, there are many corporate laboratories that can be engaged by DoD. In general, then DoD should strive to rapidly adopt leading-edge technologies developed by industry and try to minimize developing custom technologies for DoD use. (See Recommendation 5.15.)

PRINCIPLE 2: SIMPLIFY PROCUREMENT

An impediment to being "fast" is DoD's long and complicated procurement process. A simpler and easier procurement process would make more suppliers willing to work with DoD and would enable DoD to more quickly adopt the latest technology. To briefly summarize a complex area, DoD's procurement process needs to be simplified, using a Kaizen process or other similar formalism, so that suppliers will not have to deal with a lot of bureaucracy and thus will be more willing to be a supplier to DoD. (See Recommendation 5.9.)

PRINCIPLE 3: ESTABLISH UPGRADE SCHEDULES FOR MICROELECTRONICS SYSTEMS

One way to overcome the impediment presented by long and complicated procurement processes is for DoD to mandate a periodic and regular upgrade schedule for some of its electronic hardware that is vulnerable to obsolescence because of the rate at which commercial microelectronics advances. There may be opportunities for DoD to learn these practices from cell phone manufacturers (e.g., Apple) that adhere to regular product upgrade cycles. As one example, signal processing electronics in radar systems could be upgraded more frequently so that these systems are only one or two generations behind commercial state-of-the-art (SOTA) microelectronics. Mandating upgrade schedules would provide DoD with an incentive to keep abreast of industry's advances. Furthermore, the predictability may make industry more willing to work with DoD because of the anticipated opportunity to do business.

As detailed in the Chapter 3 discussion of modernization and subsequent recommendations, DoD should implement a regular upgrade schedule for integrated circuits to ensure that manufacturing capacity is always available and manufacturers are more incentivized to be suppliers to DoD. (See Recommendation 4.2.)

PRINCIPLE 4: COORDINATE CLOSELY WITH THE DEPARTMENT OF COMMERCE AND OTHER AGENCIES

The Microelectronics Commons (ME Commons) may be an effective way to fund research and development (R&D) on disruptive technologies for DoD use, but certain provisions are important to ensure that useful results ensue. To start with, most members on the committee with experience in developing advanced semiconductor technologies feel that a 5-year funding timeframe is very likely too short for a disruptive technology to make it all the way through the research, development, and prototyping phases to being ready for manufacturing—a process that usually takes 10 years or more. What will happen when disruptive technologies get halfway through the "valley of death" and then funding stops? If DoD and the Department of Commerce (DOC) collaborate in setting up and running new capabilities such as the ME Commons and the National Semiconductor Technology Center (NSTC), then there is a greater chance of being granted longer-term funding from Congress and achieving ultimate success in bringing innovative technologies from the research phase to manufacturing. DoD should identify down-selection criteria for new project and center proposals to better ensure that centers invite the strongest partners to participate. Another important provision is that projects have clearly defined deliverables that are measurable. And when projects are not making good progress toward their deliverables, then those projects should be terminated to free up funds for more promising projects and hubs.

The intention of ME Commons to set up core facilities (laboratories and fabrication plants—labs and fabs) is certainly important in being able to demonstrate the disruptive technologies needed by DoD, but these laboratories and fabs can be very expensive and time-consuming to set up. The high capital cost of semiconductor manufacturing facilities is the main reason why the industry is now concentrated in relatively few locations and few companies. The ME Commons should try to use existing facilities at universities, such as the Albany NanoTech Complex, the Interuniversity Microelectronics Centre (IMEC), the National Institute of Standards and Technology (NIST), and government laboratories, as a way to get quick access to processing facilities. But where new equipment or facilities are needed, ME Commons should not try to do this on its own. ME Commons should use its money to expand existing government facilities. It makes no sense to duplicate the facility efforts of NSTC and the National Advanced Packaging Manufacturing Program (NAPMP) when most of this expensive equipment can and should be used for both DoD and DOC purposes.

In pursuit of developing disruptive technologies for DoD use, it might sound like a simple matter for DoD to provide R&D funds to the appropriate institutions, whether they be commercial, academic, or start-ups and small firms. But challenges that DoD can face in getting these institutions to commit to taking on these projects fall into the following two categories: (1) commercial companies may not want to be distracted by R&D projects that are too far from their main business, and (2) academic researchers and small to medium-sized companies may be very interested in the project but do not have access to the SOTA lab/fab facilities needed to do the research and bring the technology to full manufacturing. The NSTC and NAPMP facilities being planned by DOC are likely good options for supporting disruptive technology projects at universities and small to medium-sized companies as they evolve from research to prototyping phases.

As detailed in Chapters 3, 4, and 5, DoD and DOC have a common interest in supporting the U.S. semiconductor industry and developing advanced technologies. Thus, DoD and DOC should collaborate in setting up and running the expensive R&D facilities needed to do this. (See Recommendation 4.4.)

PRINCIPLE 5: COLLABORATE WITH LEADING COMPANIES FOR CUSTOM NEEDS

Convincing a commercial company to take on a disruptive technology project should start with DoD partnering with companies that already have a technology or products that are not significantly different from the new technology or would be a technology that would be commercially interesting to them. And if the disruptive technology is expected to be a derivative version of commercial technology, then DoD should strive to be a fast follower (Box 6-1).

BOX 6-1
Fast Follower

By "fast follower," the committee means that the Department of Defense (DoD) should strive to be a leading-edge customer and the first military user in the world to rapidly adopt, and incorporate into its systems, the newest microelectronic technologies developed by commercial industry. Here the committee is considering mainstream logic, memory, and analog integrated circuit process technologies that are advanced by industry leaders such as Intel, GlobalFoundries, Micron, Texas Instruments, Analog Devices, Nvidia, Qualcomm, Advanced Micro Devices, Inc., and others. DoD cannot hope to match the capabilities of those companies with a captive, defense-unique industrial base, and should instead strive to adopt those technologies for DoD use just as they become available for commercial purposes. DoD may not be the first to ship systems using a new technology node, but it must be the first globally to deploy military systems using that technology (because of militarily advantageous features— lowest energy use, highest performance, smallest size and weight).

DoD should offer intellectual property (IP) rights that are not too restrictive for the commercial company and ensure that red tape associated with the procurement process is not overly complicated and onerous.

To summarize this point, and as discussed in earlier chapters, when DoD needs a customized process to meet its needs, it should collaborate with a leading-edge manufacturer in a way that minimizes red tape and has clear financial benefits for the manufacturer.

PRINCIPLE 6: ENSURE UNIVERSITY RESEARCHERS AND START-UPS HAVE ACCESS TO ADVANCED EQUIPMENT

Academic researchers and start-up companies may have much to offer DoD in terms of out-of-the-box thinking and an eagerness to take on revolutionary projects. But their lack of access to modern lab/fab facilities may prevent their ideas from ever reaching the prototype phase or manufacturing readiness. This is a common problem for academic researchers and start-up companies, whether the technology being researched is for DoD or commercial application. To address this need, the CHIPS and Science Act of 2022 (the CHIPS Act), under DOC/NIST direction, is proposing that NSTC provide access for these smaller institutions to a modern 300 mm wafer processing facility. Such a facility will be very expensive to set up and run and calls out for DoD and DOC to both contribute funds to operate this facility in a collaborative manner. In some cases, universities have access to process tools needed for the research, but the lack of a professional staff to operate and maintain these tools significantly limits their availability and usefulness. Thus, some money from DoD or DOC can be valuable when used to support professional staff at university facilities.

DOC's NSTC program is intended to be set up to support the advanced semiconductor process needs of universities and start-up companies. DoD (including through the ME Commons) should collaborate with DOC to ensure that NSTC can support the academic projects and start-up companies that are of potential value to DoD. (See Recommendation 5.2.)

PRINCIPLE 7: SUPPORT ACCESS TO SCALE-UP CAPITAL

When the proposed NSTC facility becomes operative, it will certainly be very helpful for start-ups and university groups that need prototyping facilities. However, most hardware innovations, such as those that would result from semiconductor start-ups, require sustained funding for much longer periods of time than private investors are willing to provide funding. In the long run, when such technologies have potential to make a difference in DoD operations, DoD should consider nurturing them by playing a role akin to a patient venture capitalist. There are existing organizations that could help with this activity. DoD and the intelligence agencies have similar aims in this area, so the Director of National Intelligence's (DNI's) In-Q-Tel could assist in this aim. The Defense Innovation Unit (DIU) and DoD's new Office of Strategic Capital could also assist. Focusing on DoD needs and its objective of sustaining an engagement with the semiconductor industry would be critical to making sure that DoD does not fall behind and yet is simultaneously contributing to supporting innovations that will keep the U.S. semiconductor industry ahead of competitors.

To summarize, DoD and DOC should both be looking for long-term sources of funding to support revolutionary integrated circuit projects that will need more than 5 years to reach manufacturing readiness. (See Recommendation 5.3.)

PRINCIPLE 8: SUPPORT THE TRANSITION FROM PROTOTYPING TO MANUFACTURING

The Flemish Government helped to establish and has continued to fund IMEC in Belgium for four decades, and the European Union (EU) has also made investments. The amount of government funding has been rather stable, with the fraction of the total budget from government sources gradually decreasing as IMEC became more successful in attracting funding from commercial companies. IMEC is a remarkable source of advanced 300 mm processing capabilities, for large and small companies, although the IMEC infrastructure is not complete enough to provide a path to manufacturing readiness. Ideally, with DOC and DoD support, NSTC could provide IMEC-like capabilities for U.S. companies doing R&D on advanced technologies. But it must be recognized that IMEC and the Albany NanoTech Complex are excellent facilities for supporting research, development, and prototyping, but not for bringing projects to full manufacturing readiness. Projects need to transition to a commercial manufacturing facility to have any reasonable hope of reaching manufacturing readiness. Only commercial manufacturers have the tools, personnel, and experience to reach these later stages.

There may be several semiconductor facilities that can be used for early R&D, but thought needs to be given up front to how technologies coming through these organizations can be transitioned to a commercial manufacturer. The sooner a commercial manufacturer is involved, the better, and government support of this transition will accelerate innovation.

PRINCIPLE 9: BUILD INTERNATIONAL COLLABORATIONS FOR ACCESS TO THE INTERNATIONAL SEMICONDUCTOR ECOSYSTEM

It might seem ideal to return to the 1960s–1970s when the United States had a dominant position in all parts of the semiconductor ecosystem: logic technology, memory technology, wafer fab equipment, semiconductor manufacturing chemicals, and innovative circuit design. However, those days are long past and highly unlikely to return. Today there is a global semiconductor ecosystem with strong players in Asia, Europe, and the United States, and the United States has benefited from being part of this global ecosystem. The CHIPS Act is a good step in strengthening the U.S. semiconductor industry, but in sectors where the United States does not have a leadership position, *multiple* suppliers from *friendly* countries should be set up. To this end, DoD, DOC, and the Department of State should collaborate on identifying and nurturing friendly countries that are home to key suppliers to the semiconductor ecosystem. The goal should be to strengthen both the United States and its global partners. DoD should also consider doing collaborative R&D projects with strong non-U.S. companies in friendly countries where no good alternative exists in the United States.

International competitors are investing heavily in their own semiconductor capabilities to build strong industries. For example, China has announced investments of some $150 billion. The European Union is investing $43 billion, India $30 billion, Japan $6.5 billion, and South Korea has announced its own significant investments, as discussed in Chapter 2. In sum, the United States is entering with sizable investments but is somewhat late to the race.

The United States needs to establish strong working relationships with semiconductor organizations in allied and friendly nations. To ensure a geographically diverse and robust integrated circuit supplier base, DoD, DOC, and the Department of State should collaborate to identify strong suppliers in friendly countries that can augment the U.S. supplier base. (See Recommendation 4.3.)

PRINCIPLE 10: STREAMLINE ENVIRONMENTAL REVIEWS

Another area of consideration for DoD concerns environmental regulations and controls that are applied to the construction of semiconductor facilities. DoD and DOC should collaborate to ensure that the National Environmental Policy Act and other environmental reviews are streamlined and do not pose a significant delay in the construction of advanced semiconductor fabs.

PRINCIPLE 11: FOSTER INFORMATION EXCHANGE

An alternative mechanism for DoD to keep abreast of cutting-edge developments in microelectronics that may significantly impact its operations that depend heavily on electronics is to work with industry to develop a framework that fosters early information exchange on leading-edge microelectronics technologies. Such information exchange has the potential to be a win–win proposition for the companies involved and for DoD. Companies acquire a willing and ready customer, and DoD gets access to state-of-the-art technology. The critical challenge is finding a workable framework that is agreeable to both parties. While it is common today for private contractors to be embedded in DoD facilities to improve efficiency of operations, the reverse process of embedding DoD employees can be more difficult. DoD may therefore have to incentivize companies to assign technical liaisons, who are full-time employees of the companies but whose special role is to look out for how cutting-edge microelectronics could benefit DoD. These individuals would be responsible for interfacing with DoD counterparts, and for making DoD aware of developments that should be considered for upgrading or replacing electronic systems.

DoD should be kept aware of leading-edge technologies developed by commercial industry by employing their own team of semiconductor technology experts who are continually in touch with industry leaders and can then make recommendations to DoD program managers in selecting the most appropriate integrated circuit technologies.

All these principles are general, with specifics captured in Chapters 4 and 5 in the form of recommendations. These principles aim to provide general guidance for DoD in its effort to assure access to the microelectronics it needs for success.

7

Workforce Development in the Semiconductor Industry

The committee's statement of task first requests the committee to examine workforce development issues related to semiconductors in the first question, where it requires a review of "barriers" to "sustainable and resilient production of semiconductors." Workforce availability is a leading barrier. In the second question, the committee is asked to review "public–private partnership strategies" that address a series of issues, including "workforce development." In the third question, the committee is asked to review unique "advantages and challenges" for the Department of Defense (DoD) in approaching additional factors, including "workforce development." In summary, the statement of task requests an examination of workforce development from three perspectives: the extent, first, that it constitutes a "barrier" to a successful semiconductor production approach; second, how a public–private partnership (PPP) approach could enable a workforce development strategy; and third, what challenges and advantages DoD faces in pursuing a PPP effort for workforce development.

There are two parts to the issues around workforce development in the semiconductor sector: needs that can be met through reforms to higher education for scientists and engineers for semiconductor research, development, and fabrication, and needs that require reforms in workforce education for the technical workforce at semiconductor facilities. Each is addressed below, preceded by a discussion of projected semiconductor sector employment needs and followed by a discussion of DoD's potential role in such reforms. This chapter closes with a discussion of the improving access to foreign-born scientists and engineers.

WORKFORCE DEVELOPMENT GAPS IN
THE SEMICONDUCTOR SECTOR

The United States faces an overall talent gap in its technical workforce of which the gap in semiconductor fields is one part. Based on Bureau of Labor Statistics employment projection data, a study by the Semiconductor Industry Association and Oxford Economics projects rapidly growing demand in U.S. jobs requiring proficiency in technical fields with these jobs more than doubling within this decade. The study projects that by 2030 there will be a gap of more than 100,000 technicians, more than 270,000 engineers, and approximately 1 million computer scientists in filling jobs across all advanced technology industries in the United States.[1] This derives from a problem of insufficient numbers of U.S. students pursuing science, technology, engineering, and mathematics (STEM) fields. This is exacerbated by the fact that many STEM majors pursue nontechnical occupations outside STEM fields. Offsetting these problems, U.S. higher education institutions attract significant numbers of STEM students from abroad, particularly at the graduate level, including 50 percent of master's engineering graduates and 60 percent of engineering PhD graduates.[2] However, some 80 percent of international master's STEM students leave the United States, as do 25 percent of the STEM PhDs. These overall problems in the U.S. STEM talent pool means trouble for the semiconductor talent pool.

The semiconductor industry in 2023 directly employs, according to industry data, approximately 345,000 in the United States, including 206,000 in manufacturing semiconductor chips, 30,000 in semiconductor machinery manufacturing, 100,000 in design, and 9,000 in developing the specialized tools used in design.[3] Three-quarters of this total is in technical work, the remainder in supporting roles (management, administrative, sales, logistics, etc.). This technical workforce falls into three broad categories: semiconductor *technicians*, who operate and maintain equipment used in making semiconductor chips and components; semiconductor *engineers*, who research, develop, and design, or improve semiconductor devices and fabrication processes; and semiconductor *computer scientists*, who design and develop software and hardware for semiconductor systems and technologies primarily within the chip design area. Shortfalls are projected in each of these three fields.[4]

---

[1] Semiconductor Industry Association (SIA) and Oxford Economics, 2023, *Chipping Away, Assessing and Addressing the Labor Market Gap Facing the U.S. Semiconductor Industry*, July 10, https://www.semiconductors.org/wp-content/uploads/2023/07/SIA_July2023_ChippingAway_website.pdf.

[2] SIA and Oxford, 2023, *Chipping Away*, pp. 9–10.

[3] See SIA and Oxford, 2023, *Chipping Away*, p. 11. These totals are for the direct semiconductor workforce, not for indirect positions in the supply chain. If such indirect employment is considered, the direct and indirect workforce is approximately 1.9 million.

[4] SIA and Oxford, 2023, *Chipping Away*, p. 12.

An analysis by McKinsey projects that the U.S. semiconductor industry output will grow significantly, with revenues growing from the current $575 billion to $1 trillion by 2030.[5] Given this projected growth, analysts project that there will be job openings for 134,300 technicians (both new jobs and replacements for workers retiring or leaving), 69,000 engineers, and 34,500 computer scientists.[6] In turn, using the U.S. Census Bureau's American Community Survey data, the analysis predicts a talent shortfall in filling these positions of 67,000, or 58 percent of the new jobs across semiconductor manufacturing and design. This includes 13,400 computer scientists, 27,300 engineers (at PhD, master's, and bachelor's levels), and 26,400 technicians. The study was completed prior to passage of the CHIPS and Science Act of 2022 (the CHIPS Act) and its corresponding investments, so may understate talent demand in this sector. Furthermore, the semiconductor sector's talent gaps are only a portion of the overall STEM workforce shortage noted above. A business-as-usual approach to workforce education in the United States is clearly going to be inadequate for meeting the projected need in both STEM-based positions overall and semiconductor sector positions in particular. Change is required, given the level of the problem.

How accurate are these employment projections, drawn from the Bureau of Labor Statistics and Census Bureau and industry data? No projections 6 years ahead can be fully accurate. The semiconductor industry often experiences up-and-down cycles; most recently, chip sales went into a slump in 2023 after record levels in 2021 and 2022. However, given the increasing pervasiveness of chips in products throughout the economy, the corresponding projected industry revenue, and the major capital investments stimulated in the United States from the CHIPS Act, totaling more than $200 billion as of the end of 2022,[7] the longer-term projections appear to be plausible.

WORKFORCE DEVELOPMENT CHALLENGES IN HIGHER EDUCATION

A major part of industry's projected gap in the semiconductor sector, as noted, is for computer scientists (13,400) and semiconductor engineers (27,300, including 5,100 PhDs, 12,300 master's, and 9,900 bachelor's degree holders). These categories require higher education.[8]

---

[5] O. Burkacky, J. Dragon, and N. Lehmann, 2022, "The Semiconductor Decade: A Trillion-Dollar Industry," McKinsey & Company, https://www.mckinsey.com/industries/ semiconductors/our-insights/the-semiconductor-decade-a-trillion-dollar-industry.

[6] SIA and Oxford, 2023, *Chipping Away*, p. 14.

[7] R. Casanova, 2022, "The CHIPS Act Has Already Sparked $200 Billion in Private Investments for U.S. Semiconductor Production," SIA, December 14, https://www.semiconductors.org/the-chips-act-has-already-sparked-200-billion-in-private-investments-for-u-s-semiconductor-production.

[8] SIA and Oxford, 2023, *Chipping Away*, p. 4.

Although it is a worldwide trend, undergraduate enrollment in the physical sciences and engineering has been in decline in the United States in favor of majors such as computer science. The semiconductor sector requires both. Bachelor's degrees awarded by U.S. higher education institutions in academic year 2020–2021 for computer science majors totaled 104,874, while physical science majors totaled 29,238 and mathematics majors totaled 89,398.[9] Bachelor's degree numbers in turn drive the numbers for follow-on graduate degrees. Although computer science degrees are rising, they are still only 4 percent of all bachelor's degrees awarded.[10]

BEST PRACTICE EXAMPLES FOR PROFESSIONAL WORKFORCE EDUCATION

While computer science enrollments have been growing in recent years, the curriculum is usually focused on software and typically does not include semiconductor design or fabrication. This means that computer science graduates have no preparation (and therefore may have no particular interest) in working in the semiconductor hardware sector.

Borivoje Nikolić and colleagues at University of California, Berkeley (UC Berkeley) provide an example of a targeted effort to try to overcome this problem.[11] They offer a hardware-focused computer architecture course to sophomores majoring in computer science to attract these majors to hardware. The aim is to get students designing chips in 14 to 15 weeks using a simplified design system they have developed, called Chipyard. Two follow-on hardware design courses are offered, as well as internships with companies, which are crucial to nurturing interest. Chipyard is a design system that includes configurable, open-source, generator-based designs that can be used across multiple hardware development stages, which was initially funded by the Defense Advanced Research Projects Agency (DARPA) and now the National Science Foundation (NSF). This effort requires considerable resources to sustain, with two engineers to maintain it and a system administrator to keep design programs updated, and to manage the licensing for design tools, and provision of the necessary computer infrastructure. Intel has supported the program, and its pilot packaging line is available for the UC Berkeley program to utilize. Carnegie Mellon University has been working on a comparable

---

[9] National Center for Education Statistics (NCES), Digest of Education Statistics, Table 322.10, Bachelor's Degrees by Field of Study, https://nces.ed.gov/programs/digest/d22/tables/dt22_322.10.asp?current=yes.

[10] C. Martinez and E. Pearson, 2018, "The Decline of Computer Science: Two Decades of Trends Reshape the Industry, Stepping Blocks Research," https://blog.steppingblocks.com/the-decline-of-computer-science.

[11] B. Nikolic, 2023, "Thoughts on Public–Private Partnerships to Strengthen Semiconductor Design," presentation to the committee, August 15.

program with Apple, as have several other schools. The key to Chipyard's success, which has attracted some 300 students to date, according to Professor Nikolić, is to expose students early in their computer science major to hardware and design, coupled with industry internships. To yield the PhD numbers later-on, it is necessary to target undergraduate students. Programs such as these encourage a pool of students to consider semiconductor hardware careers. UC Berkeley and Carnegie Mellon University are not alone; other efforts include those at Rochester Institute of Technology, Georgia Institute of Technology, and SUNY Albany, which have both design courses and laboratories where students can build what they design.

A group of faculty at the Massachusetts Institute of Technology (MIT) has been struggling with the same problem of trying to promote student interest in semiconductor fabrication and design. Their study[12] concluded the following:

> We need to come together through system-oriented multidisciplinary subjects, hands-on lab courses, research experiences, design exercises using modern computer-aided design (CAD) tools, well-crafted internship programs in industry, and support from industry mentors to attract students back to our disciplines. Research on pedagogy should explore new teaching methodologies that substantially shorten the long learning curve and reduce the high barrier for technology access that sits on the way to fulfilling project and internship experiences. Implementing these initiatives will require sizable investments in research and educational facilities and in staff support.

There is a distinction between the computer science degrees generally discussed above and computer engineering degrees. While computer science is largely focused now on software and programming, computer engineers couple electrical engineering principles with computer architecture, operating systems, networking, and coding so that they are trained to design, build, and test hardware components and systems. However, computer engineering attracts relatively small numbers of students compared to computer science, and access to up-to-date design software and equipment can be limiting. To address the latter problem, more student engagement with industry (e.g., via industry-informed, project-based learning and internships) would be beneficial, and CHIPS Act program R&D programs, such as the National Semiconductor Technology Center (NSTC), could support this.

Today, many semiconductor professionals come from specific science and engineering backgrounds, such as physics, chemistry, and electrical engineering. There is an opportunity for new multidisciplinary science and engineering programs that cover material across various disciplines, such as nanotechnology, artificial intelligence, and machine learning. Such multidisciplinary degrees may assist in

---

[12] J.A. del Alamo, D.A. Antoniadis, R.G. Atkins, M.A. Baldo, et al., 2021, "Reasserting U.S. Leadership in Microelectronics—A White Paper on the Role of Universities," Massachusetts Institute of Technology, https://usmicroelectronics.mit.edu/education-and-workforce-development.

meeting semiconductor talent needs, by tying training programs to fields that are viewed as more topical.

Overall, innovative education approaches will be needed in higher education to create the talent pool required for a strong semiconductor professional workforce, requiring focus by participating universities and support for appropriate personnel and infrastructure.

There are several lessons here. New curriculum offerings are needed in chip hardware and design that can be inserted into both computer science and computer and electrical engineering programs. Online courses could provide a valuable scaling mechanism if used in blended learning settings at universities with hand-on design tools such as UC Berkeley's Chipyard. Coupling such courses with industry engagement and internships is a vital element. Agreements with industry to fabricate and package student designs is another important element. In general, programs that provide access to up-to-date industry facilities for hands-on experience could be significant for both computer engineering and computer science. New multidisciplinary programs in both computer science and engineering may offer major advantages for nurturing semiconductor sector talent.

WORKFORCE DEVELOPMENT CHALLENGES FOR THE TECHNICAL WORKFORCE

Technical workers make up approximately 40 percent of the U.S. semiconductor workforce, with more than 26,000 new technician jobs projected to be required by 2030. Unfortunately, the U.S. workforce education system has many challenges.[13] The education system is disconnected from the workplace—that is, there is a major school-to-work barrier. The Department of Labor's training programs do not attain higher technical skills or help incumbent workers acquire them. In turn, the Department of Education's programs target college, not workforce, education and do not necessarily mesh with Department of Labor programs. The vocational education system was largely dismantled starting in the 1970s on the assumption that everyone would attend college, which did not prove to be the case. Community colleges, which could provide advanced training in emerging fields, are often underfunded, and their degree completion rates—often around one-third—are low, because many students return to the workforce after completing a few courses. Many colleges and universities are not linked to the other participants in the ecosystem. A system for lifelong learning is missing. Strong information systems make for sound markets, but the information system behind the U.S. labor market is weak—employers do not recognize the skills of job applicants, employees do not

---

[13] W.B. Bonvillian and S.E. Sarma, 2021, *Workforce Education, A New Roadmap*, Cambridge, MA: MIT Press, pp. 43–58, 238–242.

know what skills employers want, and educators do not know for what skills they should educate. Overall, the existing workforce education system operates at too local and at too small a scale to meet growing national demand.

BEST OVERALL PRACTICES FOR TECHNICAL WORKFORCE EDUCATION

A series of best practices to overcome these problems for the technical workforce in general are emerging for employers, community colleges,[14] and other system actors.

Opportunities for 2-Year Institutions

- *Offer short programs.* Programs focused on technical skills should typically run for 10 to 20 weeks. Workers with families and existing work obligations do not have the time needed for 2- or 4-year degrees—they need short courses.
- *Adopt credentialing.* Programs should provide certificates for specific groups of related skills, based on demonstrated competencies. These should be able to be credited (or "stacked") toward college associate and bachelor's degrees, which remain the most broadly recognized credentials.
- *Leverage online teaching.* Online modules will be critical if workforce education is going to scale to meet upcoming demand because it provides new flexibility for today's students. Online education cannot replace effective in-person instruction or hands-on work with actual equipment, but it can supplement these by conveying foundational information behind the skills and be used for some assessments.
- *Support apprenticeships.* Programs to break down the work/learn barrier should be linked to industry connecting training to actual workplaces. Community colleges can cooperate with industry on apprenticeships, as well as internships and cooperative programs, that can move students into the workplace so they can earn while building skills. These also enable students to make direct connections between the competencies they must learn and new job opportunities.

---

[14] Bonvillian and Sarma, 2021, *Workforce Education*, pp. 251–255; W.B. Bonvillian, 2021, "America Needs a New Workforce Education System," *Issues in Science and Technology*, March 9, https://issues.org/workforce-education-innovation-community-colleges-bonvillain; G. Westerman, W. Bonvillian, Z. Clochard-Bopssuet, L. Amrutha Killada, and J. Liu, 2021, "Benchmarking Phase 1 Report," in *MassBridge: Massachusetts Advanced Manufacturing Workforce Education Program Report*, September 24, https://openlearning.mit.edu/sites/default/files/inline-files/Benchmarking%20Phase%201%20Report%20%28Final%29.pdf; S. Nelson, W.B. Bonvillian, and G. Westerman, 2022, "Community Colleges' Role in Building Apprenticeships," in *MassBridge: Massachusetts Advanced Manufacturing Workforce Education Program Report*, June 30, https://mitili.mit.edu/research/community-colleges-role-building-apprenticeships.

- *Embed industry-recognized credentials into educational programs.* Many employers want an assurance of skill knowledge that a credential approved and accepted by industry provides. Credentials create an additional and parallel pathway to help students toward employment. They also ensure that academic programs are relevant to actual industry needs.
- *Develop working alliances with universities for advanced curricula and online offerings.* Universities, working with industry, can help develop curricula in emerging and advanced fields, working with community colleges. Universities can also offer online course development in concert with community colleges.

Opportunities for Industry

- *Organize groups of employers.* Single employers embarking on workforce education face varying demands over time for workers, which rise and fall with business cycles and orders. They typically cannot sustain the steady flow of students for sustainable education programs. Instead, employers should band together, including large companies and suppliers, to create and communicate a steady demand signal required for sustainable education efforts.
- *Ally with community colleges.* Groups of employers in turn should create alliances with area community colleges for the stable and sustainable skills development programs that both employers and community colleges require. They should work with community colleges to adopt the approaches listed above, as well as to help develop and implement curricula, including in advanced areas, that companies require.
- *Promote apprenticeships.* Employers are central to apprenticeship programs (as well as to paid internships and cooperative programs), which can help employers find the workers with the skills they want. Alliances with community colleges on apprenticeships can assure continuity between the on-the-job training and education programs.

These general practices could be applied for semiconductor employers and community colleges in their regions to improve semiconductor technical workforce development.

EXAMPLES OF BEST PRACTICES IN SEMICONDUCTOR TECHNICAL WORKFORCE EDUCATION

These best practices for workforce education at the technical level are highly applicable to the semiconductor sector. There have been recent examples in the semiconductor field of attempts to pursue some of these practices.

The Quick Start Program in Arizona

Arizona is host to six Intel fabrication plants (fabs), including two new ones under construction that will manufacture its most advanced chips and require 3,000 new workers. TSMC is building two fabs in Arizona to produce its 3 and 5 nm chips and expects to employ 4,500 workers. A workforce education challenge faces both companies in Arizona, and the campuses in the Maricopa Community Colleges system will play a role in meeting it. One of those campuses, Mesa, has formed an Arizona Advanced Manufacturing Institute that offers onsite employer recruitment and internships and apprenticeships, a career navigator system for career planning, and industry certifications embedded into its courses.[15] It is particularly focused on the rapidly growing semiconductor sector.

Mesa and two other Maricopa community colleges have developed a "quick start" 10-day program designed with industry to introduce students to semiconductor sector manufacturing careers (a branch of the program also serves the aerospace industry).[16] The program is highly accelerated and features adjunct expert faculty from Intel teaching basic hands-on skills, The college has acquired $1.2 million in equipment and supplies for students to work with. Students earn a certificate that qualifies them for entry level positions, and the college hosts frequent hiring fairs with industry. Students receive a $270 stipend at the end of the program to offset their costs in attending. As of October 2023, 712 students had completed certificates from the program, 65 percent were students of color and 32 percent female, 44 percent were ages 18 to 29 (most of the older workers were already in the workforce), and half were first-generation immigrants. The institute also offers a follow-on Automated Industrial Technology (AIT) program, with AIT industry certification and associate degree credit. These longer programs qualify students in areas such as robotics, power electronics, and modern maintenance. Sixty-two of the students who completed the quick-start program have returned for this additional course work.

The community colleges are working to develop better data on industry employment and retention, but an initial survey indicates that 31 percent of those who have completed the program have found semiconductor sector jobs.[17] Why are the hiring numbers not higher? Construction delays have pushed out the completion

---

[15] Mesa Community College, a Maricopa Community College, Arizona Advanced Manufacturing Institute, https://www.mesacc.edu/workforce-development/azami.

[16] L. Palmer, 2023, "Partnerships—Building a Pipeline for Scope and Scale," presentation to the committee, October 24. See also, Maricopa Community Colleges, quick start semiconductor program, https://info.maricopacorporate.com/semiconductor. The program is given at three Maricopa system community colleges, Chandler-Gilber, Estrella Mountain, and Mesa Community Colleges.

[17] J. Zinkula, 2024 "The Chip Industry's False Promise," January 9, https://www.businessinsider.com/semiconductor-chips-jobs-hiring-arizona-tech-phoenix-tsmc-intel-china-2024-1.

times for the new semiconductor fabs, and the industry is going through a downturn after expansion in 2022. However, a quick-start, intensive, short course with hands-on experience with industry still appears to offer a model that could help attract new students to semiconductor work when demand picks up.

Community College Collaboration in Ohio

Intel is building two new fabs in Ohio that will employ 3,000 people in the Columbus area and has been closely collaborating with community colleges as well as state universities to develop a workforce in a region that has not had prior semiconductor employment. Intel expects that at least 3,000 supplier jobs will develop as well.[18] Aside from constructing the new fabs, Intel has committed $100 million in Ohio, which has been matched by $50 million from NSF, for scholarships and education development. Intel developed a five-part program for workforce development, including curriculum development, faculty training, laboratory equipment, research programs on chip fabrication, and hands-on experience for students.

While Ohio's community colleges are independent, they have a history of cooperation. Working with the Ohio Association of Community Colleges and its member schools, Intel has helped develop a 1-year semiconductor certificate program, with five of the community colleges getting ready to launch this curriculum. Eleven schools plan to offer the program. The curriculum development effort was supported by Intel through a series of projects, with seven different Ohio institutions leading each, including Ohio State University and Lorain County Community College, one of the nation's leading community colleges, which is leading the "train the trainers" program to educate new faculty for the certificate program. These seven technology center projects were competitively awarded based on requests for proposals and have received $17.7 million in Intel funding through 2023.[19]

The new certificate will offer credit toward degrees as well as an industry certification recognized by Intel. Students completing the new certificate will be eligible for entry positions at the Intel fabs. Intel is making training tools to help ready the faculty in specific semiconductor skills; it will also take new faculty to its Arizona fabs to learn its equipment and work environment. Fab-sized equipment is not part of the program, because it would be too expensive for the schools to operate and maintain, but Intel is providing laboratory-scale equipment for training. Online education is part of the curriculum approach, using videos with blended

---

[18] S. Venkataramani, 2023, presentation to the committee, November 28.

[19] Intel, 2023, "Intel Invests in Ohio Education," Intel, September 24, https://download.intel.com/newsroom/2023/corporate/OH-Workforce-fact-sheet.pdf.

learning. Intel is also working with the schools to develop summer programs with hands-on training opportunities. It is also considering apprenticeship programs in cooperation with the participating Ohio schools, a number of which already have extensive "earn and learn" programs for students. Intel expects to train thousands of students through the program over the next 3 years.

To summarize, Intel, working closely with Ohio community colleges and universities, has incorporated many of the best practices listed above. The collaborative effort between the Ohio community colleges, universities, and Intel suggests how major workforce education programs at the technician level can be effectively organized. Overall, both examples show the importance of close collaboration between industry and community colleges in developing semiconductor technical workforce education. And co-location with industry of hands-on education can be particularly constructive.

MECHANISMS FOR SUPPORT OF WORKFORCE EDUCATION FOR THE DEPARTMENT OF DEFENSE TO CONSIDER

There is a series of possible mechanisms for DoD to potentially work through concerning workforce development for technician and scientist or engineer levels for the semiconductor sector. One is the network of Manufacturing USA advanced manufacturing institutes. DoD's ManTech program has provided funding support for eight of those institutes, and up to three new institutes in semiconductor fields were approved following the CHIPS and Science Act of 2022 (the CHIPS Act), sponsored by the National Institute of Standards and Technology (NIST). All such institutes have workforce education projects as part of their core mission, aimed primarily at bringing advanced manufacturing skills to the technical workforce.[20] DoD's Industrial Base and Sustainability (IBAS) program also supports workforce development for DoD's nearer-term industrial base needs, including a program to support development of a new workforce category of "technologist" to move talented technical workers into positions managing the new advanced manufacturing systems starting to be implemented in U.S. factories.[21] In addition, NSTC, supported by NIST and the Microelectronics Commons (ME Commons), formed by DoD (both discussed earlier in this report), also provide potential support mechanisms for workforce education efforts particularly at the scientist and engineering level. Finally, NSF has funding under the CHIPS Act for semiconductor education programs.

---

[20] W.B. Bonvillian, 2022, "The Playbook, for Workforce Education at Manufacturing Innovation Institutes," Report for DoD Manufacturing Technology Program, January.

[21] J. Liu and W.B. Bonvillian, 2024, "The Technologist," *Issues in Science and Technology*, Winter, https://issues.org/issue/40-2.

NSF's Advanced Technological Education (ATE) program has long-led efforts for development of advanced technology curricula and education programs at community colleges.[22]

> Finding 7.1: The semiconductor sector faces a significant talent shortage for both its professional scientist and engineering workforce and its technical workforce. NIST, NSF, and DoD all have existing workforce education programs relevant to the semiconductor sector that, with close coordination with industry and education institution efforts, can be leveraged to meet both professional and technical workforce needs for the semiconductor sector generally and for DoD-specific needs.

> **Recommendation 7.1: The Department of Defense (through the director of the CHIPS Coordination Cell) should partner with the National Institute of Standards and Technology and the National Science Foundation to implement and expand workforce education programs for the semiconductor sector, support the development of training programs, apprenticeships, and industry-recognized credentials, build alliances with community colleges, and promote regional public–private partnerships.**

For example, the advanced manufacturing institutes and the ME Commons supported by DoD might play important roles. Programs should support both the development and scaling of semiconductor training, to include the following: short programs, credentialing, online education, apprenticeships, and industry-recognized credentials. Efforts to coordinate groups of employers in support of workforce development, building alliances with community colleges, and promoting apprenticeships would be beneficial. PPPs at regional and local levels might be utilized to develop needed collaborations between semiconductor firms and suppliers, universities and community colleges, and state and local governments. DoD should be an active promoter and participant in these diverse activities.

ACCESS TO FOREIGN-BORN SCIENTISTS AND ENGINEERS

The United States has historically relied on attracting foreign-born science and engineering talent to meet its workforce needs at the professional level. As noted above, 50 percent of master's engineering graduates and 60 percent of engineering PhD graduates at U.S. universities are from abroad.[23] However, as noted, some 80

---

[22] See, for example, concerning ATE's programs, NSF, Advanced Technological Education (ATE), Program Solicitation NSF 21-598, October 14, 2021, through October 5, 2023, https://www.nsf.gov/pubs/2021/nsf21598/nsf21598.htm.

[23] SIA and Oxford, 2023, *Chipping Away*, pp. 9–10.

percent of international master's STEM students leave the United States, as do 25 percent of the STEM PhDs. Retaining more of these graduates could help significantly in meeting workforce needs in the semiconductor sector.

The immigration system limits the ability to access and retain outstanding foreign science and engineering talent trained in semiconductor fields. This situation calls out for improvement. In 2022, the U.S. Citizenship and Immigration Service (USCIS) made modest changes in its guidance for two visa categories available for STEM workers.[24] There were changes in O-1A temporary visa for "aliens of extraordinary ability" that can pave the way for "green cards" to be issued, to allow lawful permanent residence and work in the United States without full citizenship. There were also changes in the EB-2 employment visa that allows green cards for those with advanced STEM degrees. As a result, in 2022, O-1A visas increased by 30 percent to 4,570, and STEM EB-2 visas increased in 2022 by 55 percent over 2021 levels to 70,240. Those higher levels continued in 2023. However, a 1990 law still allows only 140,000 employment-based green cards to be allowed each year, with no more than 7 percent to the citizens of any particular country.

As recommended in 2022 by the President's Council of Advisors on Science and Technology (PCAST), premium processing of petitions for advanced-degree immigrants seeking to work in microelectronics, through available statutes and regulations, could help with some of this concern.[25] Legislation is needed to revise the immigration laws to reflect the reality of technology needs, including that country of origin should not matter when immigrants are selected based on their skills and education, and enabling an accelerated path to citizenship. Meanwhile, without such comprehensive reform, an intermediate step is to award lawful permanent resident status for those with advanced STEM degrees under current law outside of the country or worldwide numerical caps, for individuals working in semiconductor and other advanced technology areas. DoD could encourage USCIS to prioritize such approaches, given the importance of semiconductors and other advanced technologies to national security.

Finding 7.2: The semiconductor professional workforce in the United States depends on international students, yet there are significant regulatory immigration barriers to ensuring they are available to work in the United States.

---

[24] J. Mervis, 2023, "New U.S. Immigration Rules Spur More Visa Approvals for STEM Workers," *Science* December 27, https://www.science.org/content/article/new-u-s-immigration-rules-spur-more-visa-approvals-stem-workers.

[25] President's Council of Advisors on Science and Technology (PCAST), 2022, "Report to the President on Revitalizing the Semiconductor Ecosystem," https://www.whitehouse.gov/wp-content/uploads/2022/09/PCAST_Semiconductors-Report_Sep2022.pdf.

Recommendation 7.2: The Department of Defense (through the director of the CHIPS Coordination Cell) should advocate for reforms to the immigration system to support the semiconductor workforce, including granting lawful permanent resident status to individuals with advanced science, technology, engineering, and mathematics degrees working in the semiconductor industry.

8

Recommendations

This chapter lists all the recommendations in the report, arranged under the general topics of modernization, research, design, manufacturing, regulatory concerns, intellectual property, workforce development, and agency coordination.

MODERNIZATION

The Department of Defense (DoD) (through the Office of the Under Secretary of Defense for Acquisition and Sustainment), as part of DoD's ongoing or future public–private partnerships, should develop new processes and practices to accelerate the deployment of advanced chips into existing DoD platforms and equipment, to deploy modern electronic design automation tools, incorporating artificial intelligence, to facilitate and streamline the replacement of older chipsets thereby addressing security risks in legacy chips while improving size, weight, and performance where possible. (Recommendation 4.2)

The Department of Defense (DoD) (especially the Chief Digital and Artificial Intelligence Office) should prioritize partnering with industry over creating custom solutions, utilizing commercial technology wherever possible. DoD should partner closely with U.S. companies to creatively and nimbly adopt emerging technologies for defense purposes, with immediate urgent attention to artificial intelligence and potential superintelligent systems. (Recommendation 5.15)

The Department of Defense (through the Office of the Under Secretary of Defense for Acquisition and Sustainment, in partnership with the service branches) should develop an overarching microelectronics strategy for research, development, procurement, sustainment, and modernization. (Recommendation 5.16)

RESEARCH

The Department of Defense (through the director of the CHIPS Coordination Cell) should coordinate with the National Institute of Standards and Technology to invest in new metrology technologies that support next-generation semiconductor manufacturing. (Recommendation 4.7)

The Department of Defense (through the Office of the Under Secretary of Defense for Research and Engineering) should continue support for State-of-the-Art Heterogeneous Integrated Packaging (SHIP), SHIP 2.0, and other advanced semiconductor packaging research and development programs and initiatives, collaborating with commercial manufacturers to develop customized processes or capabilities when needed. (Recommendation 4.9)

The Department of Defense (through the Office of the Under Secretary of Defense for Research and Engineering) should focus on long-term scientific research in disruptive semiconductor technologies, including post-CMOS technologies, and ensure broader access to prototyping facilities for academic researchers and small to medium-sized firms. (Recommendation 5.3)

The Department of Defense (DoD) (through the Office of the Under Secretary of Defense for Research and Engineering) should create a flagship unit with expertise in hardware cybersecurity and modern chip design that DoD program managers can access as needed. DoD should explore embedding design teams from this unit within companies developing leading-edge chip designs to increase DoD's technical expertise. (Recommendation 5.7)

DESIGN

The Department of Defense (DoD) (through the Office of the Under Secretary of Defense for Acquisition and Sustainment) should launch an effort that crosses service branches and programs to facilitate the cost-efficient use of electronic design automation tool licenses and establish a clearinghouse for DoD custom chip designs to be reused. (Recommendation 4.5)

The Department of Defense (through the Office of the Under Secretary of Defense for Research and Engineering) should organize a significant initiative, in partnership with industry, to leverage artificial intelligence and machine learning tools to substantially reduce the time and cost of application-specific integrated circuit design and software development for defense needs. (Recommendation 5.5)

The Department of Defense (through the Office of the Under Secretary of Defense for Research and Engineering) should build skills internally for modern chip design, partner with commercial-sector experts designing chips at advanced nodes, and tap into the Department of Commerce–supported supply chains for manufacturing necessary custom chips. (Recommendation 5.6)

MANUFACTURING

The Department of Defense (DoD) Office of the Under Secretary of Defense for Research and Engineering should strengthen public–private partnerships (PPPs) that support semiconductor manufacturing, particularly for technology readiness levels 4–6. DoD should ensure these PPPs have long-range funding, agreed-upon intellectual property terms, clear goals, and success metrics. (Recommendation 4.1)

The Department of Defense (led by the Defense Logistics Agency in partnership with the director of the CHIPS Coordination Cell) should coordinate with the Departments of Energy, Commerce, and State to identify choke-point materials for semiconductor manufacturing and partner with private industry to actively cultivate a robust, geographically and geopolitically diverse supply base for each. (Recommendation 4.3)

The Department of Defense (through the Strategic Radiation-Hardened Electronics Council in coordination with the director of the CHIPS Coordination Cell) should partner with the Department of Commerce to establish a national center of excellence for radiation-hardened microelectronics design and testing to greatly accelerate the timeline for development of such components for national security and commercial applications. (Recommendation 4.6)

The Department of Defense (DoD) (through the Office of the Under Secretary of Defense for Acquisition and Sustainment) should review and update policies that limit it from manufacturing in commercial facilities. DoD should also increase the use of domestically manufactured chips where possible and simplify procurement processes to streamline access to semiconductor suppliers. (Recommendation 5.9)

The Department of Defense (DoD) (through the director of the CHIPS Coordination Cell) should act to secure longer-term federal funding support for the semiconductor sector, beyond the term of the CHIPS and Science Act of 2022, to ensure DoD has continual access to the most advanced semiconductor technologies in the world. DoD should also ensure that these technologies can be sourced from diverse foundries located in the United States and friendly nations. (Recommendation 5.12)

The Department of Defense (DoD) (through the director of the CHIPS Coordination Cell) should act to ensure that federal capital and financing supports for the semiconductor sector are tied to agreements that guarantee it access to advanced semiconductors. The DoD Office of Strategic Capital should also prioritize financing for critical technology scale-up activities. (Recommendation 5.13)

REGULATORY CONCERNS

The Department of Defense (DoD) (through the Office of the Under Secretary of Defense for Acquisition and Sustainment) should accelerate efforts to implement an evidence-based assurance system to ensure fast and secure access by DoD programs to advanced commercial semiconductors, and also simplify bureaucratic processes and update relevant DoD instructions and policies in support of this effort. (Recommendation 5.8)

The Department of Defense (DoD) (through the Defense Technology Security Administration) should collaborate with relevant organizations to reduce the administrative burdens and improve the timeliness of decisions related to International Traffic in Arms Regulations, Export Administration Regulations, and the National Environmental Policy Act. DoD should consider forming a task force to review and reform these regulations. (Recommendation 5.10)

The Department of Defense (DoD) (through the Office of the Under Secretary of Defense for Acquisition and Sustainment) should lower barriers to utilizing domestic manufacturing entities, minimizing complex bureaucracy, regulations, and requirements, to obtain the custom chips that DoD needs. The process for determining where DoD's manufacturing can be performed should have short response times so as to avoid unduly hindering DoD's technology programs. (Recommendation 5.14)

INTELLECTUAL PROPERTY

The Department of Defense (DoD) (through the Office of the Under Secretary of Defense for Research and Engineering) should manage intellectual property (IP) rights based on technology readiness levels and intended end use, recognize private ownership of IP in the majority of cases, reserve government ownership only for extreme situations, conduct an audit to track DoD's existing IP rights, and create a central, searchable records system. (Recommendation 5.11)

WORKFORCE DEVELOPMENT

The Department of Defense (through the director of the CHIPS Coordination Cell) should partner with the National Institute of Standards and Technology and the National Science Foundation to implement and expand workforce education programs for the semiconductor sector, support the development of training programs, apprenticeships, and industry-recognized credentials, build alliances with community colleges, and promote regional public–private partnerships. (Recommendation 7.1)

The Department of Defense (through the director of the CHIPS Coordination Cell) should advocate for reforms to the immigration system to support the semiconductor workforce, including granting lawful permanent resident status to individuals with advanced science, technology, engineering, and mathematics degrees working in the semiconductor industry. (Recommendation 7.2)

AGENCY COORDINATION

The Department of Defense (through the director of the CHIPS Coordination Cell) should ensure that its semiconductor research and development efforts are closely integrated with the related agendas funded through the Departments of Commerce and Energy and the National Science Foundation. (Recommendation 4.4)

The Department of Defense (DoD) (through the director of the CHIPS Coordination Cell) should closely coordinate with the Department of Commerce on CHIPS Act incentive decisions to ensure funds are directed toward firms that can meet DoD's needs. (Recommendation 4.8)

The Department of Defense (through the director of the CHIPS Coordination Cell) should coordinate with the National Institute of Standards and Technology on its semiconductor-related public–private partnership efforts, including via the Microelectronics Commons, to avoid duplication and ensure the exchange of promising advances. (Recommendation 5.1)

The Department of Defense (through the Office of the Under Secretary of Defense for Research and Engineering and the Office of the Under Secretary of Defense for Acquisition and Sustainment) should actively leverage its own programs and partner with the National Institute of Standards and Technology to connect academic researchers with industry partners and advanced fabrication and packaging facilities. (Recommendation 5.2)

The Department of Defense (DoD) (through the Chief Digital and Artificial Intelligence Office) should swiftly incorporate U.S.-origin commercially developed artificial intelligence (AI) models, chips, and infrastructure, and develop custom AI technologies for defense purposes, to maintain leadership in this transformative technology. DoD should also track leading-edge AI technologies and align DoD efforts with commercial interests to mitigate supply chain risks. (Recommendation 5.4)

Appendixes

A

Statement of Task

A National Academies of Sciences, Engineering, and Medicine–appointed ad hoc committee will identify, explore, and assess public–private partnership models that have the potential to enable assured access for the production of semiconductors in the United States. The committee will produce a consensus report that addresses the following questions.

What is the competitive position of the United States in the global semiconductor ecosystem?

The committee will examine barriers to sustainable and resilient production of semiconductors in the United States and explore what helps drive production and create reliable supply chains of materials, equipment, components, and expertise. This could include an exploration of the industrial policies of other nations in support of industries in similar critical technology sectors.

How can public–private partnerships strengthen semiconductor manufacturing?

Given the inherent strengths and weaknesses within the global microelectronics industry, the committee will explore how to tailor public–private partnership strategies to address different aspects of the supply chain, such as tool manufacturing, fabless design (without fabrication plants), electronic design automation, software development, manufacturing capability and capacity, workforce development, domestic research and engineering capture (e.g., hardware start-ups), and raw materials (e.g., wafers, rare earths). This may include an analysis of establishing a semiconductor manufacturing corporation to leverage private-sector technical, managerial, and investment expertise, as well as private capital.

When partnering with the private sector for semiconductor production, what unique challenges and opportunities exist for the Department of Defense to support sustainability and resilience in the U.S. semiconductor ecosystem?

The committee will examine unique challenges for the Department of Defense in engaging with public–private partnerships on semiconductors. In order to provide meaningful new insights into the advantages and challenges of public–private partnerships, the committee will consider issues such as the research–design–production feedback loop, intellectual property, workforce development, export controls, global marketplace considerations, and specialized Department of Defense technologies not supported by the commercial industry.

What policies for public–private partnerships could be adopted to accelerate the development and adoption of disruptive technologies in the United States that benefit the Department of Defense and dual-use needs?

Given previously described barriers and challenges, the committee will discuss and recommend approaches for the Department of Defense to drive change. The committee will conduct an assessment of, and response to, the industrial policies of other nations to support industries in similar critical technology sectors, which will include analyses and recommendations for the consideration of U.S.-international partnerships in support of the global marketplace, as well as onshoring efforts. The committee will also examine and describe resources (amounts and types of funding) and actions that could help achieve or maintain a global leadership position for each aspect of the supply chain and effectively leverage private-sector investment.

B

Public Meeting Presentations

JUNE 6, 2023

Dev Shenoy, Principal Director for Microelectronics, Office of the Under
Secretary of Defense for Research and Engineering

JUNE 27, 2023

Eric Lin, Deputy Director, CHIPS Research and Development Office

JULY 18, 2023

Lode Lauwers, Senior Vice President, Business Development and Strategy

JULY 25, 2023

**Economic and Historical Views,
Including Lessons Learned from SEMATECH**

Dan Hutcheson, Vice Chair, TechInsights

Innovation Policy

Sujai Shivakumar, Director and Senior Fellow, Renewing American Innovation
Project, Center for Strategic and International Studies

Outlooks and Analysis of the Semiconductor Ecosystem

Boston Consulting Group—Ramiro Palma, Managing Director and Partner
(Austin); and Thomas Lopez, Principal
Deloitte—Mark LaViolette, Managing Director, Supply Chains & Network
Operations, Government and Public Services; Duncan Stewart, Director,
Research for the Technology, Media, and Telecommunications Industry for
Deloitte Canada
Christie Simons, U.S. Audit and Assurance Technology, Media, and
Telecommunications Industry Leader
Tyler Schmidt, Senior Director, Strategic Business Development Group, Intel
Federal LLC

JULY 26, 2023

Department of Defense Views, Including Microelectronics Commons

Victoria Coleman, United States Air Force Chief Scientist
Antonio de la Serna, Senior Director, Microelectronics EDA Strategy, Siemens
Government Technologies and Secretary of the National Defense Industrial
Association Electronics Division

AUGUST 1, 2023

Vivek Menon, Systems Engineering Directorate, National Reconnaissance Office
Paul Schaum, Director, Office of Contracts Policy, National Reconnaissance
Office

AUGUST 8, 2023

Chris Miller, Associate Professor, The Flether School, Tufts University

AUGUST 15, 2023

Bora Nikolic, Professor, University of California, Berkeley

SEPTEMBER 5, 2023

Mukesh Khare, General Manager IBM Semiconductors; Vice President, Hybrid
   Cloud, IBM

SEPTEMBER 12, 2023

Arun Seraphin, Executive Director, Emerging Technologies Institute, National
   Defense Industry Association

SEPTEMBER 19, 2023

Mark Rosker, Director, DARPA Microsystems Technology Office
Robert Atkinson, President, Information Technology and Innovation Foundation
Alison Smith, Microelectronics Commons Technical Director, Naval Surface
   Warfare Center (NSWC)—Crane Division
Bryan Smith, Microelectronics Commons Program Manager, NSWC Crane
Jason Rathje, Director, Office of Strategic Capital, Department of Defense
Doug Fuller, Associate Professor, Copenhagen Business

SEPTEMBER 26, 2023

Dan Armbrust, Co-Founder and Director of Silicon Catalyst and Former
   President and CEO of SEMATECH

OCTOBER 3, 2023

Erik Hadland, Director, Technology Policy at the Semiconductor Industry
   Association (SIA)

OCTOBER 10, 2023

Dev Shenoy, Principal Director for Microelectronics, Office of the Under
   Secretary of Defense for Research and Engineering

OCTOBER 17, 2023

Christina Lomasney, Director, Commercialization, Technology, Deployment and
   Outreach, Pacific Northwest National Laboratory

OCTOBER 24, 2023

James Li, Senior Director, Electronics Systems Microelectronics, BAE Systems
Ezra Hall, Aerospace and Defense Business, Senior Director, GlobalFoundries
Patty Chang-Chien, Vice President and General Manager, Boeing Research &
    Technology
Mike Burkland, Senior Principal Systems Engineer, Raytheon Missile Systems

OCTOBER 25, 2023

Leah Palmer, Executive Director, AzAMI Workforce, Applied Science, Mesa
    Community College, Maricopa Community Colleges
Vanessa Peña, Senior Policy Manager, Office of Technology Transitions,
    Department of Energy
Brent Segal, Director of Technology Collaboration with the Corporate
    Technology Office, Lockheed Martin

NOVEMBER 28, 2023

Sowmya Venkataramani, Program Director, University Research and
    Collaboration Group in Intel Laboratories and Program Director, Intel
    Semiconductor Education and Research Program for Ohio

DECEMBER 5, 2023

Neil Thompson, Director, FutureTech Research Project, Massachusetts Institute
    of Technology (MIT) Computer Science and Artificial Intelligence
    Laboratory and MIT Sloan School of Management and Group Lead &
    Principal Investigator, Technologies that Create Prosperity, MIT Initiative on
    the Digital Economy

DECEMBER 12, 2023

Melinda Reed, Director for System Security in the Office of the Under Secretary
    of Defense for Research and Engineering S&T Program Protection Office

JANUARY 9, 2024

Tyler Schmidt, Senior Director of the Strategic Business Development Group
    at Intel Federal

Terri Wetteland, Senior Director of Supply Chain M&A at Intel Federal
 Procurement & Intel Foundry Services Supply Chain Capabilities

JANUARY 23, 2024

Jayson McDonald, Microelectronics Task Lead, SAIC and Contractor Support
 to PD, Microelectronics
Jon Rolf, Director, National Information Assurance Partnership, National
 Security Agency

FEBRUARY 20, 2024

Robert McFarland, Technical SETA to the Principal Director for Trusted AI and
 Autonomy in the Office of the Assistant Secretary of Defense for Critical
 Technologies

C

A Semiconductor Public–Private Partnership Playbook

Given the considerations outlined in Chapter 4, Table C-1 describes a multiphase approach that could be followed to facilitate the creation of a public–private partnership to advance semiconductor manufacturing.

TABLE C-1 Multiphase Approach for Creation of a Public–Private Partnership (PPP) to Advance Semiconductor Manufacturing

	Task	Implementation	Key Consideration(s)
Phase I (3–6 months)	Establish a single entity (e.g., an LLC) empowered to operate the PPP	Founding membership	Membership solicitation (notice of funding opportunity, request for proposal, etc.).
			This entity should act as a "single window" through which PPP members contract (rather than subcontracting with each other).
	Within this entity, create a board of directors, a Department of Defense (DoD) interface center, a business office, and advisory committees	Management board	This board should provide overall guidance on PPP objectives, budget, and spending, and evaluate progress.
		DoD PPP interface office	This DoD office should be tasked aligning the PPP's activities with DoD's mission needs.
		Central business office	This office should be tasked with daily PPP operations, including facility management, personnel, and information technology.
		Advisory committees	*Finance committee:* Determine terms and conditions of PPP membership levels, evaluate expenses, and manage budgets and contracts.
			Business committee: Serve as the business development/marketing arm of the PPP.
			Technology committee: Develop technical research agenda, goals, metrics for success, and risks.
			Defense user committee: Gather customer needs and ensure PPP activities are meeting the requirements of the Combatant Commands, major commands, and other users. Could also include defense industrial base systems integrators that will adopt microelectronics onto defense platforms.
			Intellectual property (IP) committee: Establish mutually agreeable terms and conditions for IP treatment.

continued

TABLE C-1 Continued

	Task	Implementation	Key Consideration(s)
Phase II (3–6 months)	Develop technology roadmaps, timeline, goals, and metrics for success	Establish technical working groups	These technical working groups should focus on specific subtasks, assigning timelines, goals, and risks.
	Determine partnerships necessary to execute roadmap	Procure and prepare physical space for PPP research operations (e.g., cleanrooms, warehousing), establish subcontracts with key vendors/suppliers	This should be coordinated with the finance and business committees.
Phase III (2 years)	Operations	Technical working groups	Working groups should report progress and risks to the technology committee on a monthly and quarterly basis.
		IP review	As progress is made, the IP committee should review inventions and technical advances, and classify inventions for patent applications or as a trade secret.
		Review cost, schedule, and performance	The advisory committees should convene quarterly to assess overall progress toward stated metrics of success.
	External Engagement	External advisory group	A group of representatives from non-PPP members should convene to review the PPP's progress on its stated research agenda annually.
		Coordination with associations	To every extent practical, updates on the PPP's research agenda and progress should be shared with relevant national and international industry associations, including via conferences.
		Additional PPP membership	The PPP should remain open to "late entrants" who wish to participate after the PPP has begun its initial work. Solicitation of these new members and management of their terms of membership should be handled by the business and finance committees.

D

Summary Description of the CHIPS and Science Act of 2022

Description of the Act drawn from HR 4346, 117th Congress, 2nd Sess., July 2022 (bill text), https://www.commerce.senate.gov/services/files/CFC99CC6-CE84-4B1A-8BBF-8D2E84BD7965; U.S. Congress, Senate Committee on Commerce Science and Transportation, CHIPS and Science Act of 2022, Division A Summary – CHIPS and ORAN Investment, July 2022, https://www.commerce.senate.gov/services/files/2699CE4B-51A5-4082-9CED-4B6CD912BBC8; U.S. Congress, Senate Committee on Commerce Science and Transportation. Chips and Science Act of 2022, Section-by-Section Summary, July 2022, https://www.commerce.senate.gov/services/files/1201E1CA-73CB-44BB-ADEB-E69634DA9BB9.

Key provisions of the act included the following:

- *Manufacturing grants and incentives:* $39 billion for financial assistance to construct, expand, or modernize domestic facilities and equipment for semiconductor fabrication, assembly, testing, advanced packaging, or research and development, including $2 billion specifically for legacy semiconductors. Within the incentive program, up to $6 billion may be used for the cost of direct loans and loan guarantees.
- *Advanced manufacturing investment tax credit (ITC):* In addition to the grant program, the ITC provides a 25 percent investment tax credit for investments in semiconductor manufacturing. The credit covers both manufacturing equipment as well as the construction of semiconductor manufacturing facilities. It also includes incentives for the manufacturing of the specialized tooling equipment required in the semiconductor

manufacturing process. The ITC aims to erase the difference between U.S. and foreign subsidy regimes, and, when paired with the CHIPS grant funding, seeks to reduce the up to 40 percent cost difference for leading-edge semiconductor production.

- *Research and development (R&D):* $11 billion is for R&D, which will fund a series of program elements.
- *National Semiconductor Technology Center (NSTC):* A public–private partnership to conduct advanced semiconductor manufacturing R&D and prototyping, invest in new technologies, and expand workforce training and development opportunities. Consortia of industry and universities will compete for the center award.
- *National Advanced Packaging Manufacturing Program:* A federal R&D program to strengthen advanced assembly, testing, and advanced packaging capabilities.
- *Manufacturing USA Semiconductor Institutes:* Manufacturing innovation institutes between government, industry, and academia to develop improvements of semiconductor machinery, develop advanced packaging capabilities, and design and disseminate training.
- *Microelectronics Metrology R&D:* A National Institute of Standards and Technology (NIST) research program for advanced semiconductor measurement science, standards, material characterization, instrumentation, testing, and manufacturing capabilities.
- *CHIPS for America Defense Fund* with $2 billion for the Department of Defense and the Defense Advanced Research Projects Agency to implement the Microelectronics Commons, a national network for onshore, university-based prototyping, lab-to-fab transition of semiconductor technologies, including defense-unique applications, and semiconductor workforce training.
- *Workforce education:* $200 million for education and training of the domestic semiconductor workforce, including both the engineering and technical workforce, which face near-term labor shortages, to be led by the National Science Foundation and its education division.
- *International collaboration:* $500 million for the Department of State, in coordination with the Agency for International Development, the Export-Import Bank, and the U.S. International Development Finance Corporation, to support international information and communications technology security and semiconductor supply chain activities, including supporting the development and adoption of secure and trusted telecommunications technologies, semiconductors, and other emerging technologies. The legislation also directs the Secretary of Commerce to

report on the capabilities of the U.S. industrial base to support national
defense needs in light of the global nature of the supply chain and significant interdependencies between the U.S. industrial base and that of foreign
countries in microelectronics.

NOTE: These are authorized programs but not all are being funded or fully funded
as of this writing.

E

Committee Member Biographical Information

LIESL FOLKS, *Chair*, is the vice president for semiconductor strategy and a professor of electrical and computer engineering at the University of Arizona. Previously, she performed fundamental research and development on nanoscale magnetic materials and devices in support of the data storage industry for 16 years in California's Silicon Valley at the IBM Almaden Research Center, the Hitachi San Jose Research Center, Hitachi GST Advanced Development, and Western Digital. Dr. Folks is an internationally recognized expert in the fields of magnetic materials and devices, nanoscale metrology, and spin-electronic devices. Her research and inventions have centered on (1) dynamics in magnetic materials, (2) magnetic devices for data storage and random-access memory applications, and (3) semiconductor devices for application as magnetic field sensors. She holds 14 U.S. patents and authored more than 60 peer-reviewed journal articles resulting in more than 12,000 citations. She has long been a leader in expanding inclusion in science, technology, engineering, and mathematics (STEM). She holds a PhD in physics from the University of Western Australia and an MBA from Cornell University. She served as the president of the Institute of Electrical and Electronics Engineers (IEEE) Magnetics Society from 2013–2014 and co-chaired the congressionally mandated 2020 Review of the National Nanotechnology Initiative for the National Academies of Sciences, Engineering, and Medicine. She is a fellow of the National Academy of Inventors.

MARK T. BOHR retired from Intel Corporation as an Intel senior fellow in February 2019. He was rehired by Intel as a part-time consultant in March 2021 and continued to work in that capacity until he retired in February 2024. He originally joined Intel in 1978 and worked as an integrated circuit process development engineer in the Logic Technology Development (LTD) group for his entire career. While a member of LTD, he worked on the development of 20 generations of logic and memory technologies up to Intel's 5 nm generation logic technology. He is an IEEE fellow, recipient of the 2003 IEEE Andrew S. Grove Award, recipient of the 2012 IEEE Jun-ichi Nishizawa Medal, and a member of the National Academy of Engineering. He holds 98 patents in integrated circuit processing and has authored or co-authored 51 published papers. Mr. Bohr received his BS in industrial engineering in 1976 and MS in electrical engineering in 1978, both from the University of Illinois at Urbana-Champaign.

WILLIAM B. BONVILLIAN is a lecturer at the Massachusetts Institute of Technology (MIT) in its Science Technology and Society Department, teaching courses on innovation and science and technology policy. He is a senior director for special projects at MIT's Office of Open Learning, conducting research projects on workforce education and innovation projects. Previously, he directed MIT's Washington Office, coordinating MIT's work on national research and development (R&D) and technology policy issues with R&D agencies and Congress. Prior to MIT, he served for more than 15 years as a senior policy advisor in the U.S. Senate working on innovation issues. He has testified before Congress and written extensively on science and technology policy issues in numerous journals. He is the co-author of five books on technology policy, advanced manufacturing, the Defense Advanced Research Projects Agency (DARPA), and workforce education. He is on the National Academies' National Materials and Manufacturing Board, its standing committee for the Science Policy Forum, and has served on eight other National Academies' committees. Mr. Bonvillian chaired the American Association for the Advancement of Science (AAAS) standing Committee on Science, Engineering and Public Policy between 2017 and 2021, is a member of the Babbage Forum on industrial innovation policy at Cambridge University, is on the Government Accountability Office's Polaris Council on Science and Technology, and is on the board of the Information Technology and Innovation Foundation. He has degrees from Columbia University, Yale University, and Columbia Law School.

PATRICIA CAMPBELL is a professor at the University of Maryland Carey School of Law, where she serves as the director of the Intellectual Property Law Program and the director of the Maryland Intellectual Property Legal Resource Center. Her research interests include trademark law and counterfeiting, technology policy, history of technology, and patent law, including university ownership and commercialization of inventions. Ms. Campbell recently completed a Counterfeit Microelectronics Policy Analysis for the Defense Microelectronics Activity, Department of Defense (DoD), as part of a larger Machine Vision Pilot Program in collaboration with researchers from the School of Engineering, University of Maryland, College Park. She presented on policy considerations regarding counterfeit electronic parts at the National Science Foundation (NSF) Workshop on Enterprise Network Models for Counterfeit Supply Chains in 2021 and at the CALCE/SMTA Symposium on Counterfeit Parts and Materials in 2020. Her work has been published in the *William & Mary Business Law Review*, the *North Carolina Journal of Law & Technology*, and others. Ms. Campbell received her BA in history from Carnegie Mellon University and her MA in history from the University of Pittsburgh. She then received her JD from the University of Pittsburgh School of Law in 1989 and her LLM in intellectual property law from the Santa Clara University School of Law in 2004.

WILLIAM CHAPPELL is the vice president of missions systems and the chief technology officer for the Strategic Missions and Technology Division (SMT) of Microsoft. Within the SMT team, he has developed the Azure Space and Spectrum team, as well as the strategic modeling and simulation, which includes silicon cloud design. Prior to Microsoft, he was the director of the Microsystems Technology Office (MTO) for DARPA of DoD. Serving in this position, he focused the office on three key thrusts important to national security. These thrusts included ensuring unfettered use of the electromagnetic spectrum, building an alternative business model for acquiring advanced DoD electronics that feature built-in trust, and developing circuit architectures for next-generation machine learning. The MTO creates the microelectromechanical systems, photonic, and electronic components needed to gracefully bridge the divide between the physical world in which we live and the digital realm where our information resides. Under Dr. Chappell's leadership, MTO developed the basic underpinnings of computation and sensing needed for an effective, information-driven society. Prior to his work in government, Dr. Chappell was a professor at Purdue University specializing in electromagnetics. He received his BS, MS, and PhD in electrical engineering, all from the University of Michigan.

KENNETH FLAMM is a professor emeritus (formerly Professor and Dean Rusk Chair) at the LBJ School of Public Affairs, The University of Texas at Austin. From 1993 to 1995, Dr. Flamm served as Principal Deputy Assistant Secretary of Defense for Economic Security and Special Assistant to the Deputy Secretary of Defense for Dual Use Technology Policy. Dr. Flamm, an applied microeconomist, is best known for his work on the economics of technological innovation and competition in high-tech industries. He has written numerous widely cited articles and books on international competition in high technology industries. One current research topic is the effect of a slowdown in semiconductor manufacturing innovation on industrial structure and downstream user industries, and broader strategic implications for national policy. Dr. Flamm holds a PhD in economics from MIT and a BA (with distinction) in economics (honors) from Stanford University. He has been a member of several other National Academies' committees and panels, including as the vice chair of the Panel on Comparative Innovation Policy and a member of its Science, Technology, and Economic Policy Board. Dr. Flamm currently runs an economic advisory service specializing in the use of modern data science tools to analyze real-world technology industry data and has provided economic analysis to semiconductor company clients, including AMD, Micron, Texas Instruments, and the Taiwan Semiconductor Manufacturing Company in the past.

KATHLEEN N. KINGSCOTT is currently a senior strategic advisor at the Alliance of Professionals and Consultants, Inc., in support of IBM Research. She recently retired as the vice president, strategic partnerships, IBM Research, where she was responsible for developing collaborative research partnerships among IBM, industry, academia, and government. Ms. Kingscott served as IBM's alternate member of the Semiconductor Industry Association board of directors. Previously, Ms. Kingscott held the IBM Industry Chair at the Industrial College of the Armed Forces, National Defense University. Earlier roles include the director of worldwide innovation policy for the IBM Corporation, responsible for worldwide public policy matters regarding innovation, science, and technology. Her global team provided political and legislative support on innovation policy matters ranging from fundamental and applied multidisciplinary research to semiconductor and supercomputing technology policy. She also focused on innovation-based regional economic growth. She is a founder and served as the co-chair of the Task Force on American Innovation, a coalition of more than 60 companies, university and trade associations, and professional societies that supports federal investment in scientific research. Ms. Kingscott led IBM's policy work in developing the Trusted Foundry, a partnership between industry and government to develop specialized semiconductors for defense applications. She is a member of the Innovation Policy Forum of the National Academies. Ms. Kingscott is a member of the board of managers of the American Institute of Physics Publishing.

BHAVYA LAL is the former associate administrator for technology, policy, and strategy within the office of the NASA Administrator, and responsible for providing evidence-driven advice to NASA leadership on internal and external policy issues, strategic planning, and technology investments. At the direction of the NASA Administrator, she also created and provided executive leadership and direction to the Office of Technology, Policy, and Strategy. Dr. Lal also simultaneously served as the acting chief technologist of NASA and was the first woman to hold the position in NASA's history. Prior to her associate administrator role and in the first 100 days of the Biden administration, Dr. Lal was the acting Chief of Staff at NASA and directed the agency's transition under the administration of President Biden. Before arriving at NASA, she was a member of the Presidential Transition Agency

Review Teams for both NASA and DoD. For 15 years prior to that, Dr. Lal led strategy, technology assessment, and policy studies and analyses at the Institute for Defense Analyses (IDA) Science and Technology Policy Institute for the White House Office of Science and Technology Policy, the National Space Council, NASA, DoD, and other federal departments and agencies. Before coming to IDA, Dr. Lal was the director of the Center for Science and Technology Policy Studies at Abt Associates, a global policy research and consulting firm. She is an active member of the space technology and policy community, having chaired, co-chaired, or served on six high-impact National Academies' committees. She served two consecutive terms on the National Oceanic and Atmospheric Administration Federal Advisory Committee on Commercial Remote Sensing, was an external council member of the NASA Innovative Advanced Concepts Program, and was selected to join the NASA Technology, Innovation and Engineering Advisory Committee. She co-founded and co-chaired the policy track of the American Nuclear Society's annual conference on Nuclear and Emerging Technologies in Space and co-organized a seminar series on space history and policy with the Smithsonian National Air and Space Museum. She guest lectures at universities across the country, including MIT, Georgetown University, Arizona State University, and others, and has testified multiple times to Congress and the National Space Council. Dr. Lal's analyses have been at the center of almost all space-relevant policies for the last decade. For her outstanding contributions to the development of astronautics, she was nominated and selected to be a member "academician" of the International Academy of Astronautics. Dr. Lal holds bachelor's and master's degrees in nuclear engineering from MIT, a second master's from MIT's Technology and Policy Program, and a PhD in public policy and public administration from The George Washington University. She is a member of both the nuclear engineering and public policy and public administration honor societies and has published more than 50 papers in peer-reviewed journals and conference proceedings.

OMKARAM (OM) NALAMASU is the senior vice president and the chief technology officer of Applied Materials, Inc. He leads the company's R&D, innovation, value-added strategic partnerships with global academia, research institutes, customers, supply chain partners, and government funding agencies. Prior to joining Applied Materials, Dr. Nalamasu served as an NYSTAR Distinguished Professor of materials science and engineering at Rensselaer Polytechnic Institute, where he also served as the vice president of research. Prior to that, he held key R&D leadership positions at AT&T Bell Laboratories, Bell Laboratories/Lucent Technologies, and Agere Systems, Inc. Dr. Nalamasu has made seminal contributions to the fields of optical lithography and polymeric materials science and technology. He is an IEEE fellow and has received numerous awards; authored more than 180 papers, review articles, and books; and holds more than 120 worldwide issued patents.

He serves on several national and international advisory boards, including on the boards of The Tech Interactive in San Jose, Global Semiconductor Association, Global Corporate Venture Leadership Society, and Singapore's MTC IAP. He received his PhD from The University of British Columbia, Vancouver, Canada, in 1986 in chemistry. He was elected to the National Academy of Engineering and is a board member of the National Academies' National Materials and Manufacturing Board. Dr. Nalamasu is also the president of Applied Ventures, LLC, the venture capital fund of Applied Materials, and serves on the board of directors of the Global Semiconductor Alliance.

ELIAS TOWE is the Albert and Ethel Grobstein Professor at Carnegie Mellon University where he teaches in the Departments of Electrical and Computer Engineering and Materials Science and Engineering. Dr. Towe's group pursues research in optical and quantum phenomena in semiconducting materials for applications in novel photonic devices and systems that enable a new generation of information processing systems for communication, computing, and sensing. His current research has been on heterogeneous integration of III-V semiconducting compounds on silicon and transition-metal dichalcogenides on silicon. His group has also been working on quantum computing, communications, and sensing. Dr. Towe was educated at MIT where he received his SB, SM, and PhD from the Department of Electrical Engineering and Computer Science, where he was a Viton Hayes fellow. He is a fellow of IEEE, the American Physical Society, OPTICA (formally OSA), and AAAS.

JOHN VERWEY is an East Asia national security advisor at Pacific Northwest National Laboratory (PNNL). Before joining PNNL, Mr. VerWey worked for the U.S. Trade Representative, where he served as a director for investment and staff liaison to the Committee on Foreign Investment in the United States. He has also served as a semiconductor industry analyst at the U.S. International Trade Commission (USITC) and an industry analyst at the Department of Commerce. In these roles Mr. VerWey led investigations for the executive branch and various congressional committees on economic competitiveness issues and co-led a multiyear assessment of the microelectronics defense industrial base. He has authored 15 reports on the microelectronics industry in recent years, including work published by USITC, the *Journal of International Commerce and Economics*, *IEEE-Computer*, and Georgetown University's Center for Security and Emerging Technology. He holds a graduate degree in international political economy from the London School of Economics and undergraduate degrees in Asian studies and history from Gonzaga University. Mr. VerWey has published articles and social media posts on semiconductor supply chains and industry analysis, CHIPS Act implementation, and Chinese semiconductor industrial policy.

F

Disclosure of Unavoidable Conflicts of Interest

The conflict of interest policy of the National Academies of Sciences, Engineering, and Medicine (http://www.nationalacademies.org/coi) prohibits the appointment of an individual to a committee authoring a Consensus Study Report if the individual has a conflict of interest that is relevant to the task to be performed. An exception to this prohibition is permitted if the National Academies determine that the conflict is unavoidable and the conflict is publicly disclosed. A determination of a conflict of interest for an individual is not an assessment of that individual's actual behavior or character or ability to act objectively despite the conflicting interest.

Mark T. Bohr has a conflict in relation to his service on the Committee on Global Microelectronics: Models for the Department of Defense in Semiconductor Public–Private Partnerships because he is a consultant for Intel and holds Intel stock. The National Academies have concluded that in order for the committee to accomplish the tasks for which it was established, its membership must include at least one member with current experience developing semiconductor technologies for microprocessors and logic devices, one of four major semiconductor product groups, at a leading U.S.-based semiconductor manufacturer. As described in his biographical summary, Mr. Bohr is a current consultant and former employee of Intel, has extensive experience in integrated circuit process development and logic and memory technologies. Mr. Bohr also has expertise regarding domestic research and engineering capture (e.g., hardware start-ups) and raw materials (e.g., wafers, rare earths), as well as the research–design–production feedback loop. The National Academies have determined that the experience and expertise of Mr. Bohr is needed for the committee to accomplish the task for which it has been established. The National Academies could not find another available individual with the equivalent experience and expertise who did not have a conflict of interest. Therefore, the National Academies have concluded that the conflict is unavoidable. The National Academies believe that Mr. Bohr can serve effectively as a member of the committee, and the committee can produce an objective report, taking into account the composition of the committee, the work to be performed, and the procedures to be followed in completing the study.

Kathleen N. Kingscott has a conflict in relation to her service on the Committee on Global Microelectronics: Models for the Department of Defense in Semiconductor Public–Private Partnerships because of consulting work for IBM Research, the research and development (R&D) division for IBM, and her ownership of IBM stock. The National Academies have concluded that in order for the committee to accomplish the tasks for which it was established, its membership must include at least one member with current experience with the operations of a leading industrial research organization, including R&D in advanced semiconductors and related public policy issues. As described in her biographical summary, Ms. Kingscott was responsible for developing collaborative research partnerships among IBM, industry, academia, and government, and she has extensive experience leading worldwide public policy matters regarding innovation, science, and technology in the semiconductor industry. In addition, Ms. Kingscott led IBM's policy work in developing the Trusted Foundry, a partnership between industry and government to develop specialized semiconductors for defense applications. The National Academies have determined that the experience and expertise of Ms. Kingscott is needed for the committee to accomplish the task for which it has been established. The National Academies could not find another available individual with the equivalent experience and expertise who did not have a conflict of interest. Therefore, the National Academies have concluded that the conflict is unavoidable. The National Academies believe that Ms. Kingscott can serve effectively as a member of the committee, and the committee can produce an objective report, taking into account the composition of the committee, the work to be performed, and the procedures to be followed in completing the study.

Omkaram (Om) Nalamasu has a conflict in relation to his service on the Committee on Global Microelectronics: Models for the Department of Defense in Semiconductor Public–Private Partnerships because of his employment at Applied Materials, Inc., and he holds Applied Materials, Inc., stock as well as stock in other companies including the Taiwan Semiconductor Manufacturing Company, Micron, Broadcom, SOXL, and Qualcomm. Dr. Nalamasu is also the president of Applied Ventures, LLC, the venture capital fund of Applied Materials, and serves on the board of directors of the Global Semiconductor Alliance. The National Academies have concluded that in order for the committee to accomplish the tasks for which it was established, its membership must include at least one member with current deep industry expertise on the manufacturing of tools and equipment needed in the early stage of the semiconductor ecosystem. As described in his biographical summary, as the senior vice president and the chief technology officer at Applied Materials, Dr. Nalamasu leads the company's R&D, innovation, and strategic partnerships. Dr. Nalamasu's has extensive experience in all phases of semiconductor manufacturing capability and capacity, including the research–design–production feedback loop, intellectual property, export controls, and in developing and maintaining public–private partnerships and global marketplace considerations. The National Academies have determined that the experience and expertise of Dr. Nalamasu is needed for the committee to accomplish the task for which it has been established. The National Academies could not find another available individual with the equivalent experience and expertise who did not have a conflict of interest. Therefore, the National Academies have concluded that the conflict is unavoidable. The National Academies believe that Dr. Nalamasu can serve effectively as a member of the committee, and the committee can produce an objective report, taking into account the composition of the committee, the work to be performed, and the procedures to be followed in completing the study.

G

Acronyms and Abbreviations

AMD	Advanced Micro Devices, Inc.
ASIC	application-specific integrated circuit
ASML	Advanced Semiconductor Materials Lithography
CEO	chief executive officer
CHIPS	Creating Helpful Incentives to Produce Semiconductors
CMOS	complementary metal–oxide–semiconductor
DARPA	Defense Advanced Research Projects Agency
DFARS	Defense Federal Acquisition Regulation Supplement
DMEA	Defense Microelectronics Activity
DOC	Department of Commerce
DoD	Department of Defense
DRAM	dynamic random access memory
DSB	Defense Science Board
EAR	Export Administration Regulations
EDA	electronic design automation
ERI	Electronic Resurgence Initiative
EUV	extreme ultraviolet

FET	field-effect transistor
GOCO	government-owned, contractor-operated
IMEC	Interuniversity Microelectronics Centre
IRDS	International Roadmap for Devices and Systems
ITAR	International Traffic in Arms Regulations
ITRI	Industrial Technology Research Institute
MIT	Massachusetts Institute of Technology
MQA	Microelectronics Quantifiable Assurance
NAPMP	National Advanced Packaging Manufacturing Program
NASA	National Aeronautics and Space Administration
NDAA	National Defense Authorization Act
NGMM	Next-Generation Microelectronics Manufacturing
NIST	National Institute of Standards and Technology
NSF	National Science Foundation
NSTC	National Semiconductor Technology Center
PPP	public–private partnership
RAMP	Rapid Assured Microelectronics Prototypes
SEMATECH	Semiconductor Manufacturing Technology
SHIP	State-of-the-Art Heterogeneous Integrated Packaging
SIA	Semiconductor Industry Association
SOTA	state-of-the-art
SRC	Semiconductor Research Corporation
STEM	science, technology, engineering, and mathematics
TRL	technology readiness level
TSMC	Taiwan Semiconductor Manufacturing Company